Springer Actuarial

Springer Actuarial Textbooks

Editors-in-Chief

Hansjoerg Albrecher, Department of Actuarial Science, University of Lausanne, Lausanne, Switzerland

Michael Sherris, School of Risk & Actuarial, UNSW Australia, Sydney, Australia

Series Editors

Daniel Bauer, Wisconsin School of Business, University of Wisconsin-Madison, Madison, USA

Stéphane Loisel, ISFA, Université Lyon 1, Lyon, France

Alexander J. McNeil, University of York, York, UK

Antoon Pelsser, Maastricht University, Maastricht, The Netherlands

Gordon Willmot, University of Waterloo, Waterloo, Canada

Hailiang Yang, Department of Statistics & Actuarial Science, The University of Hong Kong, Hong Kong, Hong Kong

This subseries of Springer Actuarial consists of graduate textbooks.

Pierre Devolder • Sébastien de Valeriola

Actuarial Pension Funding Theory

Pierre Devolder
Institute of Statistics, Biostatistics and
Actuarial Sciences
Université Catholique de Louvain
Louvain-la-Neuve, Belgium

Sébastien de Valeriola
Information and Communication Science
Department
Université libre de Bruxelles
Brussels, Belgium

ISSN 2523-3262　　　　　　　ISSN 2523-3270　(electronic)
Springer Actuarial
ISSN 2523-3300　　　　　　　ISSN 2523-3319　(electronic)
Springer Actuarial Textbooks
ISBN 978-3-031-85267-1　　　ISBN 978-3-031-85268-8　(eBook)
https://doi.org/10.1007/978-3-031-85268-8

© The Editor(s) (if applicable) and The Author(s), under exclusive license to Springer Nature Switzerland AG 2025

This work is subject to copyright. All rights are solely and exclusively licensed by the Publisher, whether the whole or part of the material is concerned, specifically the rights of translation, reprinting, reuse of illustrations, recitation, broadcasting, reproduction on microfilms or in any other physical way, and transmission or information storage and retrieval, electronic adaptation, computer software, or by similar or dissimilar methodology now known or hereafter developed.
The use of general descriptive names, registered names, trademarks, service marks, etc. in this publication does not imply, even in the absence of a specific statement, that such names are exempt from the relevant protective laws and regulations and therefore free for general use.
The publisher, the authors and the editors are safe to assume that the advice and information in this book are believed to be true and accurate at the date of publication. Neither the publisher nor the authors or the editors give a warranty, expressed or implied, with respect to the material contained herein or for any errors or omissions that may have been made. The publisher remains neutral with regard to jurisdictional claims in published maps and institutional affiliations.

This Springer imprint is published by the registered company Springer Nature Switzerland AG
The registered company address is: Gewerbestrasse 11, 6330 Cham, Switzerland

If disposing of this product, please recycle the paper.

Preface

The purpose of this book is to offer a broad and up-to-date manual on the underlying actuarial principles of pensions, covering social security schemes, private pension funds and life insurance. The topic of pensions is highly interdisciplinary, encompassing legal, actuarial, economical and even philosophical points of view. In this book, we deliberately focus on the actuarial aspects of pensions. This book is not a detailed technical or professional guide for professional actuaries in the field. Nor is it a book on the politics or economics of pensions; we wanted the contents to remain relevant for a long time and not be directly driven by the current experiences in some countries.

We have deliberately presented the basic principles in a coherent and unified manner, whether they apply to social security schemes or to private pension plans. Indeed, many existing books on that topic are in general only focused either on social security problems or on aspects of pension funds. We have tried here to present a global actuarial vision of pensions, applicable to public programs as well as to occupational pension plans. We have also included more recent techniques such as Notional Defined Contribution or hybrid pension schemes for social security programs. We want to focus on fundamental issues, using as much as possible understandable elements (for instance, in many modern books, formulas are presented in continuous time with integrals, but this is in many cases just for mathematical beauty and classical sums are enough in practice). We have also renounced to introduce stochastic elements that would ask for more theoretical knowledge. There are already more specialized books and papers on some aspects of pensions which make use of elements from stochastic finance, but that is clearly not our intention here.

The book is aimed primarily at undergraduates and graduates students, and actuaries involved in pensions, but also anyone with basic quantitative training who is interested in pension issues (mathematicians, economists, accountants, ...). The aim of the book is clearly to become a reference text for a first actuarial course on pensions. We view the book as self-contained and readable by students starting their courses in actuarial science or by anyone interested in pension issues and who has a basic algebraic knowledge. With this last point in mind, we have included an appendix summarizing the basic actuarial tools and notation that are used in the book. Simultaneously, we think that the book is novel, in that it considers the whole spectrum of pension mechanisms, including both social security mechanisms and occupational pensions. For this reason, it may also be of interest to actuaries who are less familiar, for instance, with Pay-As-You-Go actuarial elements.

This book is composed of an introductory chapter, three parts and a technical appendix. We have also included a list of notation and actuarial symbols. After a first chapter illustrating the importance of the pension issue in our society, we present in the first part the general actuarial principles of pension schemes. After a brief description of the design of pension plans and some basic elements of demography, the main methods of financing a pension scheme are described in a unified way. The second part gives a detailed analysis of the Pay-As-You-Go methodologies for social security programs. Recent concepts such as notional defined contribution plans, hybrid schemes and point systems are presented from an actuarial perspective. The third part, devoted to fully funded mechanisms, presents the classic individual and collective funding methods, such as the unit credit or aggregate cost methods, and looks at the actuarial analysis of gains and losses. The technical appendix summarizes some basic tools of mathematical finance and life contingencies used in the book.

The detailed content of this book is as follows. The first part (Chaps. 2–4) begins with a chapter aimed at providing a general view of the mechanism of a pension scheme: type of benefits, design of guarantees, method of financing or financing vehicle. The fundamental distinction between Defined Benefit schemes and Defined Contribution schemes is introduced. Chapter 3 introduces in a synthetic manner the demographic elements that will be useful in the analysis of the financial flows of schemes; fundamental notions such as the dependence ratio or the longevity risk are defined. Chapter 4 addresses in a very general manner the problem of financing a pension scheme. A classification of the major financing methods is presented and a first macroeconomic comparison between Pay-As-You-Go and fully funding is made, leading to the phenomena of the first and last generation and to what is known as the Samuelson–Aaron paradox.

The second part of this book (Chaps. 5–8) focuses on a detailed analysis of Pay-As-You-Go schemes. The majority of social security plans are Defined Benefit schemes financed by Pay-As-You-Go. Chapter 5 explains these classical arrangements, including extensions based on the creation of a buffer fund. However, we can consider Pay-As-You-Go systems not based on a Defined Benefit design. The next three chapters present recent developments in that direction. Chapter 6 is devoted to one of the most important developments of the last few decades in social security: the technique of Notional Defined Contribution schemes combining Pay-As-You-Go and Defined Contribution. In Chap. 7, we study hybrid pension plans where we try to combine Defined Benefit and Defined Contribution. In particular, we study the Musgrave rule of risk sharing between actives and retirees. Chapter 8 looks at point systems, both in a Defined Contribution and in a hybrid form.

The third part of this book (Chaps. 9–12) is devoted to actuarial methods based on fully funding. After describing the basic principles of capitalization in Chap. 9, we describe the main individual (Chap. 10) and collective (Chap. 11) funding methods. The different methods are explained in terms of cost structure (the concept of normal cost) and actuarial liabilities (the concept of accrued liabilities). Two applications are presented: the pension gap linked to mobility during a career and actuarial anticipation linked to early retirement. Finally, Chap. 12 is devoted to the analysis of gains or losses resulting from differences between ex ante assumptions and ex post realization. After outlining the origins of actuarial gains and losses, whether they are asset gaps or liability gaps, we show how different financing methods react to these phenomena.

This book reflects more than 20 years of actuarial courses on pension theory in various universities such as Louvain-la-Neuve, Brussels, Strasbourg or Rabat. Our first thanks go therefore to our students; by their questions, reactions and suggestions on these fascinating topics, they have helped us to present as best as possible the concepts. We want also to express our gratitude to our colleagues and particularly to Jennifer Alonso Garcia, Carmen Boado Penaz, Inmaculada Dominguez Fabian and Massimiliano Menzietti. Our discussions and common researches and trainings on pensions all these last years have been so exciting and crucial for the writing of this book. Finally, we thank our families for their daily support and patience!

Brussels, Belgium
April 2025

Pierre Devolder
Sébastien de Valeriola

Declarations

Competing Interests The authors have no conflicts of interest to declare that are relevant to the content of this book.

Contents

1	**The Pension Challenge**		1
	1.1	The Future of Pensions	1
	1.2	A General Taxonomy of Pension Systems	2
		1.2.1 The Sponsor	2
		1.2.2 The Affiliates	2
		1.2.3 The Definition of Benefit	3
		1.2.4 The Choice of a Funding Mechanism	3
		1.2.5 The Multi-Pillar Approach	4
	1.3	Some Pension Issues	4
		1.3.1 Social Adequacy	4
		1.3.2 Financial Sustainability	4
		1.3.3 Individual Equity	5
		1.3.4 Transparency	5
	1.4	Main Risk Factors	5
		1.4.1 Demographic Factors	5
		1.4.2 Economic Factors	7
		1.4.3 Financial Factors	8

Part I General Pension Theory

2	**Pension Schemes**		15
	2.1	Pensions and Pension Schemes	15
		2.1.1 The Bismarckian Model	18
		2.1.2 The Beveridgean Model	18
		2.1.3 Models Around the World	18
	2.2	Design of the Pension Scheme	19
		2.2.1 Defined Benefit Schemes (DB)	19
		2.2.2 Defined Contribution Schemes (DC)	20
		2.2.3 Hybrid Schemes	20
	2.3	Benefits of the Pension Scheme	21
		2.3.1 Defined Benefit Schemes	21

		2.3.2	Defined Contribution Schemes	33
	2.4	Financing the Pension Scheme		36
		2.4.1	Actuarial Equilibrium	36
		2.4.2	Fully Funded or Pay-As-You-Go Scheme?	37
3	**Introduction to Demography**			**43**
	3.1	Lexis Diagram		43
	3.2	Population Models		44
		3.2.1	A Discrete Model	44
		3.2.2	A Continuous Model	46
	3.3	Stable Populations		48
		3.3.1	Stable Discrete Populations	48
		3.3.2	Stationary Discrete Populations	49
		3.3.3	How to Obtain Stable Populations	49
		3.3.4	Stable Continuous Populations	50
		3.3.5	An Example of a Continuous Unstable Population	50
	3.4	Demographic Ratios		52
	3.5	Demographic Risks		53
		3.5.1	Longevity Risk	54
		3.5.2	Demographic Growth Risk	55
		3.5.3	A Two-Period Model	55
	3.6	Multiple-Decrement Tables		58
4	**General Funding Systems**			**61**
	4.1	General Actuarial Equilibrium Equation of a Pension Scheme		61
		4.1.1	Actuarial Equilibrium and Funding Methods	61
		4.1.2	Workers' Reserve and Retirees' Reserve	64
		4.1.3	Demographic Projections	66
	4.2	Classification of Funding Methods		67
		4.2.1	Pay-As-You-Go Method	67
		4.2.2	Scaled Premium PAYG Method	69
		4.2.3	Variant: Scaled Premium Method with Buffer Fund	70
		4.2.4	Terminal Funding PAYG Method	71
		4.2.5	Scaled Terminal Funding Method	73
		4.2.6	Fully Funded Methods	73
		4.2.7	Numerical Example	76
	4.3	Funded or Pay-As-You-Go Scheme: The Samuelson–Aaron Rule		81
		4.3.1	First and Last Generations	81
		4.3.2	Funded and Pay-As-You-Go Schemes in Stationary Phase	81
		4.3.3	Terminal Funding Schemes in Stationary Phase	85

Part II Pay-As-You-Go Pension Schemes

5 Pay-As-You-Go Social Security Schemes 89
   5.1   General Principles 89
   5.2   Internal Rate of Return of a Pay-As-You-Go Scheme 91
   5.3   A Macroeconomic Indicator 92
   5.4   Parameters of a Social Security Scheme 93
        5.4.1   Categories of Beneficiaries 94
        5.4.2   Normal Retirement Age 94
   5.5   Public Pension Reforms 95
        5.5.1   Alternative Sources of Funding 95
        5.5.2   Creation of a Buffer Fund 95
        5.5.3   Parametric Reforms 96
        5.5.4   Partial or Total Switch to Fully Funded Defined Contribution Schemes 96
        5.5.5   Notional Accounts and Point Systems 97
   5.6   Buffer Funds and Leveling Techniques 98
        5.6.1   General Principles 98
        5.6.2   The Leveling Method 98
        5.6.3   The Equalization Method 102

6 Notional Defined Contribution Schemes 105
   6.1   Basic Actuarial Mechanism 105
        6.1.1   Notional Capital 105
        6.1.2   Conversion of the Notional Capital into an Annuity 107
        6.1.3   Revaluation of the Retirement Annuity 109
   6.2   Actuarial Neutrality 109
   6.3   Notional Accounts: DC or DB? 111
   6.4   Notional Accounts and Replacement Rate 113
   6.5   Canonical Choice of Notional Accounts for a Steady-State Scheme 114
   6.6   Variants of the Canonical Choice for a Steady-State Scheme 119
   6.7   A Three-Period Non-Stationary Equilibrium Model 122
        6.7.1   Assumptions 122
        6.7.2   Notional Accounts Scheme 123
        6.7.3   Stationary Special Case 125
        6.7.4   Non-Stationary General Case 125
        6.7.5   Variants 127
   6.8   Integration of Mortality Profits 131
        6.8.1   Introduction of the Survival Dividend 131
        6.8.2   Partial Survival Dividend 133
        6.8.3   Survival Dividend and Longevity Funding 133
        6.8.4   General Multi-Age Model 135

	6.9	Implicit Debt and Pay-As-You-Go Actuarial Balance Sheet	138
		6.9.1 Short-Term or Long-Term Equilibrium	138
		6.9.2 Balance Sheet of a Closed System Scheme	140
		6.9.3 Contribution Asset and Turnover Duration	144

7 Hybrid DB/DC Systems 149
    7.1 Musgrave Rule 149
    7.2 General Risk Sharing Mechanisms 156
        7.2.1 Proportional Risk Sharing 156
        7.2.2 Convex Risk Sharing 160
    7.3 Continuous Time Model 163
    7.4 Differentiated Adjustment on Current Pensions 169
        7.4.1 Proportional Risk Sharing Between the two Retirees Generations 172
        7.4.2 Risk Entirely Borne by Older Retirees 172
        7.4.3 Risk Entirely Borne by Younger Retirees 173
        7.4.4 Proportional Risk Sharing with Guarantee 173

8 Point Systems 175
    8.1 Pay-As-You-Go DC Point Systems 175
        8.1.1 General Principles 175
        8.1.2 Standard Actuarial Equilibrium 176
        8.1.3 Actuarial Equilibrium Generation by Generation 177
        8.1.4 Contributions Call-Up Rate 178
        8.1.5 Rate of Return of Point Systems 180
    8.2 Hybrid Point Systems 181
        8.2.1 Basic Equations 181
        8.2.2 Calculation of the Value of the Point 183
        8.2.3 Age Coefficient and Normal Retirement Age 184
        8.2.4 Automatic Adaptation of the Parameters 185
        8.2.5 Automatic Adjustment of the Reference Career Duration 187

Part III Fully Funded Pension Schemes

9 General Logic of Fully Funded Pension Plans 191
    9.1 Characteristics and Classification of Funded Methods 191
        9.1.1 The Speed of Reserve Evolution 192
        9.1.2 Individual or Collective Nature 192
    9.2 Basic Quantities for Funded Methods 194

10 Individual Funding Methods 197
    10.1 General Principles of Individual Funded Methods 197
        10.1.1 First Criterion: Funding with or Without Projection 198
        10.1.2 Second Criterion: Funding by Successive Single Premiums or by Constant Premiums 198

	10.1.3	Third Criterion: Whether or Not Past Services are Funded Separately	199
	10.1.4	Crossing the Criteria	199
10.2	Unit Credit Cost		200
	10.2.1	General Principle	200
	10.2.2	Introducing a Back Service	202
10.3	Projected Unit Credit Cost		203
10.4	Individual Level Premium		204
10.5	Individual Level with Supplemental Liabilities		206
10.6	Projected Individual Level Premium		208
10.7	Projected Individual Level Percent		208
10.8	Normal Entry Age		210
10.9	Pension Gap		212
10.10	Early Retirement and Actuarial Anticipation		215

11 Collective Funding Methods ... 219
- 11.1 General Principles of Collective Funding ... 219
- 11.2 Leveling Techniques ... 220
- 11.3 Aggregate Cost ... 223
- 11.4 Attained Age Normal ... 227
- 11.5 Frozen Initial Liability ... 228

12 Actuarial Gains and Losses ... 229
- 12.1 Origin of Actuarial Gains and Losses ... 229
- 12.2 Amortization in Unit Credit Cost ... 230
  - 12.2.1 Asset Gap ... 231
  - 12.2.2 Liability Gap ... 232
  - 12.2.3 Scheme Modification ... 233
- 12.3 Amortization in Individual Level Premium ... 233
  - 12.3.1 Asset Gap ... 234
  - 12.3.2 Liability Gap ... 235
  - 12.3.3 Scheme Modification ... 235
- 12.4 Amortization in Aggregate Cost ... 236
- 12.5 Stochastic Amortization Model ... 237
  - 12.5.1 Introduction ... 237
  - 12.5.2 Stochastic Model for the Fund Value ... 238
  - 12.5.3 Asymptotic Behavior of the Fund Value and the Contributions ... 240
  - 12.5.4 Optimal Amortization Period ... 247

A Technical Appendix ... 251
- A.1 The Two Basic Tools of Actuarial Science ... 251
- A.2 Compound Interest and Capitalization ... 252
  - A.2.1 Capitalization over One Year ... 252
  - A.2.2 Capitalization over Several Years ... 252

		A.2.3	Required Capitalization Time	254
		A.2.4	Capitalization of an Annuity	254
	A.3	Discounting		256
		A.3.1	Discounting over One Year	256
		A.3.2	Discounting over Several Years	256
		A.3.3	Annuity in Present Value	257
	A.4	Life Tables		258
		A.4.1	Probability of Surviving One Year	259
		A.4.2	Probability of Death over One Year	259
		A.4.3	Probability of Surviving Several Years	261
	A.5	Life Contracts		261
		A.5.1	Pure Endowment	261
		A.5.2	Temporary Life Annuity	262
		A.5.3	Lifetime Annuity	263
		A.5.4	Application: Level Premium	264
B	**Further Reading**			**267**
Index				**271**

Notation and Symbols

L	retirement lump sum (single amount)
R	retirement pension (regular annual amount)
π	contribution rate
N	career duration
N_1	career duration – past services (back service)
P	salary cap
S	salary
S^*	projected salary
$\hat{S}$	capped salary
$\overline{S}$	average salary
x	age
x_0	initial age
x_r	retirement age
ω	ultimate age
p	premium
j	annual pension growth rate (revaluation rate)
g	annual salary growth rate
d	annual population growth rate
r	rate of return
i	discount rate
$L(x,t)$	population function (number of individuals aged x at time t)
$l(x,t)$	continuous population density at age x at time t
$N(t)$	size of the population at time t
$E(x,t)$	number of individuals entering a population at age x at time t
$V(t)$	reserves/provisions of a pension scheme at time t
$V_a(t)$	reserves for active workers
$V_r(t)$	reserves for retirees
$D(t)$	dependence ratio at time t
$v = 1/(1+i)$	discount factor
t, T	instants
$SS(t)$	total payroll at time t for a given population

$SR(t)$	total pension benefits at time t for a given population
δ	benefit ratio
ρ	continuous population growth rate
$e_x(t)$	life expectancy at age x
$p_t(x, x')$	survival probability at age x' for an individual aged x at time t
ℓ_x	number of survivals at age x for a given life table
C	contribution
F	financial assets (fund)
NC	notional capital
$N_A(t)$	number of active workers
$N_R(t)$	number of retirees
G	conversion factor
NDC	Notional Defined Contribution pension scheme
μ	risk sharing coefficient
VP	value of the point
NP	number of points
TD	turnover duration
AL	accrued liability
UAL	unfunded accrued liability
$PVFB$	present value of future benefits
$PVFS$	present value of future salaries
ADJ	adjustment contribution
rr	replacement rate
DB	Defined Benefit pension scheme
DC	Defined Contribution pension scheme
PG	pension gap
PAYG	Pay-As-You-Go
β	sustainability coefficient
VP	value of the point
NP	number of points
ac	age coefficient
ξ	call rate
N^*	reference career duration

List of Figures

Fig. 1.1	Historical evolution and projection of life expectancy at birth in the whole world (data source: United Nations)	6
Fig. 1.2	Evolution of life expectancy at 65 from 1950 to 2020 (data source: Human Mortality Database)	6
Fig. 1.3	Evolution of total fertility rate (historical and projected) from 1960 to 2060 (data source: OECD, Pensions at a glance 2021)	7
Fig. 1.4	Comparison of age pyramids in 1950 and in 2020 (data source: Human Mortality Database)	8
Fig. 1.5	Evolution of active and retirees populations from 1950 to 2020 (data source: Human Mortality Database)....................	9
Fig. 1.6	Employment rates among seniors in 2020 (data source: OECD, Pensions at a glance 2021)	10
Fig. 1.7	Comparison of the effective exit ages (for men and women) and the "normal" exit age (as defined in the law) in 2020 (data source: OECD, Pensions at a glance 2021)	10
Fig. 1.8	Evolution of inflation for OECD countries 1970 to 2023 (data source: OECD)	11
Fig. 1.9	Long term interest rate (10-year government bonds) in the Euro Zone from 1991 to 2023 (data source: OECD)................	11
Fig. 2.1	Lifeline in the Lexis diagram	37
Fig. 2.2	Lexis diagram of one fully funded risk community	38
Fig. 2.3	Lexis diagram of one Pay-As-You-Go risk community	38
Fig. 2.4	Lexis diagram of one Scaled Premium Pay-As-You-Go risk community	40
Fig. 2.5	Lexis diagram of one Terminal Funding Pay-As-You-Go risk community	40
Fig. 2.6	Lexis diagram of the Collective Funded risk community	41
Fig. 3.1	Elements of a general Lexis diagram	44
Fig. 3.2	Age pyramid	45
Fig. 7.1	Evolution of the benefit ratio in Example 7.4	169

Fig. 7.2	Evolution of the contribution rate in Example 7.4	169
Fig. 10.1	Time line	201
Fig. 10.2	Time line	211
Fig. 10.3	Pension annuity without change of employer	215
Fig. 10.4	Pension annuity with change of employer (pension gap)	215
Fig. A.1	Evolution of the capital when capitalized over several years	253
Fig. A.2	Evolution of the MR life table cohort	259
Fig. A.3	Probabilities of death q_x from the MR table	260

List of Tables

Table 2.1	Comparison between first and second pillars	17
Table 2.2	Value of α as a function of the wage growth rate	24
Table 2.3	Salary evolution	24
Table 2.4	Contribution rates obtained in Example 2.2	25
Table 2.5	Contribution rates obtained in Example 2.4	28
Table 2.6	Career profiles	34
Table 2.7	Benefits by career category	36
Table 2.8	Replacement rate by career category	36
Table 2.9	Replacement rate as a function of rate of return	36
Table 2.10	Replacement rate as a function of rate of return and salary growth rate	36
Table 2.11	Replacement rate and length of the period during which contribution is paid	36
Table 2.12	Comparison between Pay-As-You-Go and fully funded schemes	39
Table 4.1	Contribution rate of a Pay-As-You-Go scheme	68
Table 5.1	Examples of benefit ratios and Pay-As-You-Go rates of return	92
Table 5.2	Evolution of the balanced fund	101
Table 6.1	Progressive accumulation of the notional account	106
Table 6.2	Dependence of the replacement rate on various parameters in notional accounts	114
Table 6.3	Balance sheets for different types of pension schemes	141
Table 7.1	Contribution rate and benefit ratio formulas in DB, DC and DM	155
Table 7.2	Replacement rate in DB, DC and DM according to the dependency ratio	155
Table 7.3	Contribution rates in DB, DC and DM according to the dependency ratio	155

Table 7.4	Comparison of the effect of demographic risk on the benefit ratio as a function of the risk-sharing coefficient μ (initial dependency ratio: $D_1 = 0.40$)	159
Table 7.5	Comparison of the effect of demographic risk on the contribution rate as a function of the risk-sharing coefficient μ (initial dependency ratio: $D_1=0.40$)	160
Table 7.6	Contribution rate and benefit ratio obtained by varying the parameter η	162
Table 8.1	Rate of return as a function of the dependency ratio	180
Table 10.1	Individual funded methods	200
Table 10.2	Pension gap percentage as a function of n, the time of change of employer, as obtained in Example 10.1	215
Table 11.1	Overview of pure collective funding methods	220
Table 12.1	Limit value M^*	246
Table 12.2	Limit value M^{**}	250
Table A.1	Excerpt of the Belgian MR life table	258

Chapter 1
The Pension Challenge

Abstract In this preliminary chapter, we introduce some basic definitions and motivate the pension challenge, which this book aims to instill the actuarial foundations of.

1.1 The Future of Pensions

The future of our pension systems is without question one of the major social and economic challenges that our societies will meet in the coming decades. The necessity of good pension structures has emerged as a consequence of the general and universally accepted idea that everybody will stop working at a given age (retirement age) and will therefore need replacement incomes.

Pension structures concern all of us, and are more and more crucial, considering the increasing number of years people will live after retirement age. Simultaneously, the topic of pensions is generally considered as highly technical and complex. This complexity is due to the many underlying issues: political, financial, legal, actuarial and even sometimes philosophical dimensions. The importance of the challenge is also due to the inevitability of the phenomena. The threats to public pension systems are part of our future history: a lower birth rate, a reduction in the length of working careers with increasing training and, at the same time, a considerable increase in life expectancy.

In this context of profound demographic change, passionate debates are taking place between promoters of Pay-As-You-Go systems and those in favor of opening up to the financial markets through a partial transition to funded systems.

Various reforms are emerging around the world, aimed at fundamentally rethinking both the role of social security and how it works. In this context, it seems more important than ever to have at our disposal the basic technical tools to understand the issues. Broadly speaking, we can say that nearly everywhere, two fundamental kinds of pension schemes exist: either social security schemes sponsored by States or public institutions, or private regimes organized at the level of companies, professional groups and individuals. Not surprisingly the methods and techniques used for these different kinds of vehicles are quite different and often analyzed separately. However, because of their increasing interdependency, it is more and more important to have a global view and understand the common actuarial principles underlying all these pension instruments.

1.2 A General Taxonomy of Pension Systems

Many kinds of pension schemes exist all over the world and it is sometimes difficult to establish a proper classification of all these mechanisms. Situations can be quite different from one country to another, depending on culture, history, social level, political decisions, ... The basic common purpose is still the same: to guarantee to people leaving the job market at retirement age some form of income in order to survive. Technically, we can recognize a lifetime annuity.

Four main criteria can be used then in order to obtain a first global taxonomy of such pension systems:

- The sponsor: who is the "manager" of the system?
- The affiliates: who are the beneficiaries?
- The definition of benefits: how to determine the level of benefits after retirement age?
- The choice of a funding mechanism: how to finance the plan?

These four properties of a pension scheme can be seen as the ID card of the system.

1.2.1 The Sponsor

The sponsor is responsible for the the management of the pension scheme. It may have very different forms, characterized mainly by its size and its collective dimension:

- A State through a social security scheme,
- A company through an occupational pension plan,
- A professional order through a specific regime for its members (such as a lawyer, ...),
- The individual through a personal pension plan.

1.2.2 The Affiliates

The scheme has to define precisely who is entitled to receive pension benefits after retirement age. Here are some examples:

- All the citizens of a country for a social security system,
- Civil servants for a social security system,

1.2 A General Taxonomy of Pension Systems

- All the employees of a company for an occupational pension plan,
- Only the white collars of a company.

1.2.3 The Definition of Benefit

The first purpose of a pension arrangement is to obtain revenues after retirement age for their affiliates. One has therefore to define the ambition of the scheme and the way to define the pension amount for each member. This ambition can be quite different from one system to another, reflecting the proportions given to the two natural main goals of a pension scheme: enabling the elderly to avoid poverty and keeping a good standard of living after retirement.

From a technical point of view, two main systems can be defined:

- Defined Benefit scheme: a system which explicitly defines the pension that will be received at retirement age (for instance 60% of the final salary as a lifelong pension benefit),
- Defined Contribution scheme: a system which explicitly defines the contributions that must be payed during the working period; the benefits are then just the amount generated by these contributions (for instance: yearly contributions of 5% of the salaries).

1.2.4 The Choice of a Funding Mechanism

The benefits provided by any pension structure must be financed in one way or another.

A first criterion is the contributory nature of the system:

- Contributory scheme: a system where there is a link between the payment of contributions during the working life and the payment of benefits after retirement age. From an actuarial point of view, these systems use an insurance principle based on some actuarial equivalence,
- Non-contributory scheme: a system without a direct link between contributions and benefits. These systems are typically assistance programs, often financed by general tax.

In contributory systems, there are many actuarial mechanisms to finance the plan (i.e. the mathematical definition of the link between contributions and benefits). We can already highlight two basic techniques here:

- Pay-As-You-Go: a financial technique where the contributions paid by the active people are directly used to pay the benefits of the retirees at the same time,
- Fully Funding: a financial technique where the contributions paid by one cohort/ generation affects the pension payments of this cohort.

1.2.5 The Multi-Pillar Approach

Of course, in a given country, for a given individual, we can have a superposition of different schemes with different characteristics. This is sometimes referred to as the "multi-pillar" approach. For instance, an individual could be simultaneously affiliated to:

- A social security system for employees organized at a state level, based on a Defined Benefit philosophy and financed through pay as you go,
- An occupational plan for white collars in his/her company, based on a Defined Contribution system and fully funded,
- A personal pension plan in the form of an individual life insurance (fully funded by definition).

1.3 Some Pension Issues

Whatever the form and the characteristics of a system, some basic issues have to be monitored when creating, managing or reforming a pension structure.

1.3.1 Social Adequacy

The first goal of a pension structure is to provide sufficient retirement benefits. Therefore, it is crucial for the individual to check the level of benefits generated by the schemes. Of course, this can be done in a multi-pillar way, taking into account the various available pillars of pensions. In order to measure this social adequacy, we can use the concept of replacement rate or benefit ratio, comparing the level of pension benefits to salaries.

1.3.2 Financial Sustainability

Whatever the financing technique, pensions represent a cost to be paid by someone! Therefore, one must look at the future expenses generated by pension promises and check the financial viability of the scheme in the coming years. Future contribution rates to pay or other ratios such as the benefits compared to the GDP can be natural measurement tools.

1.3.3 Individual Equity

A pension plan can generate adequate benefits and remain financially sustainable, while based on unfair principles (such as for instance gender discrimination). Equity must therefore be integrated into the architecture of the scheme. Because of its long term aspect, a pension can generate intra-generational inequities but also inter-generational inequities (for instance by creating huge debts to be paid by the next generations).

1.3.4 Transparency

Pension architectures are often very difficult to catch by the affiliates, who are nevertheless very interested by their future. Complete simplicity is probably impossible to achieve; however, communication is a key aspect of a good pension strategy in order to build confidence between the sponsor and the affiliates.

1.4 Main Risk Factors

An important characteristic of a pension is its long-term aspect, requiring us to take into account many risk factors if we want to measure and analyze social adequacy and financial sustainability on a given horizon. There are various risk factors influencing the evolution of a pension system, from a demographic point of view as well as from financial or economic points of view. Actuaries have to include in their valuation all these externalities. Many data have been published about the importance of these factors and their influence on the management of pension schemes. Our aim in this section is not to be exhaustive but just to illustrate for the readers the spectacular evolution in our changing world, affecting the sustainability and adequacy of social security programs and pension funds.

1.4.1 Demographic Factors

Not surprisingly, demographic elements play a central role in the valuation of pension schemes. Over the last few decades, in many countries, the population structure has significantly evolved. Aging is nowadays a common trend, mainly induced by two factors:

- Increase of life expectancy,
- Decrease of fertility.

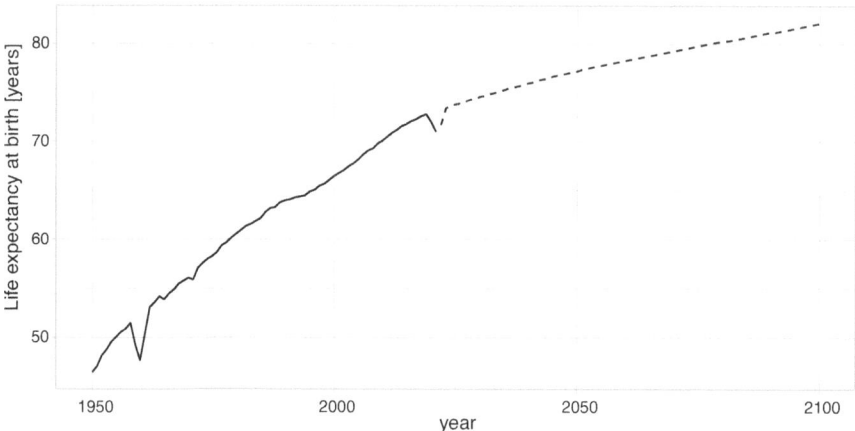

Fig. 1.1 Historical evolution and projection of life expectancy at birth in the whole world (data source: United Nations)

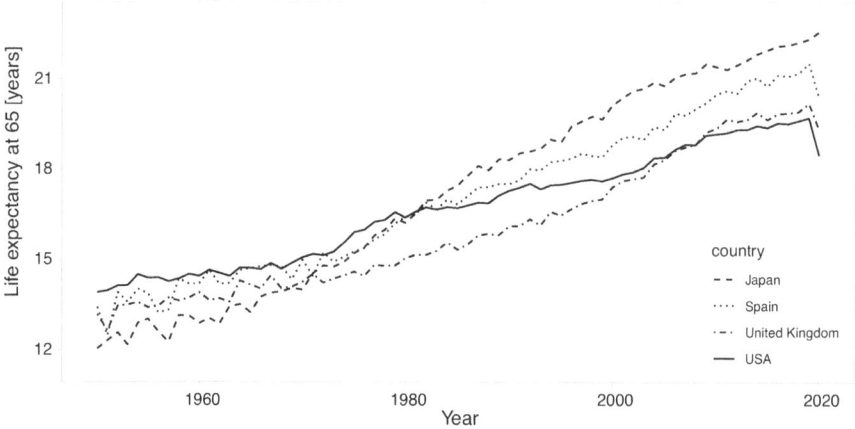

Fig. 1.2 Evolution of life expectancy at 65 from 1950 to 2020 (data source: Human Mortality Database)

Figure 1.1 shows the important worldwide increase of life expectancy at birth between 1950 and 2100 (historical and projected data).

In one century, from 1950 to 2050, life expectancy will rise from less than 50 to more than 75. For pension purposes, life expectancy at retirement age (for instance 65) is even more relevant. Figure 1.2 gives the evolution of this life expectancy at 65 for some countries. In 60 years, from 1960 till 2020, life expectancy at 65 has increased from values less than 15 to around 20.

On the other hand, the decrease of fertility is illustrated in Fig. 1.3. While the fertility rate was largely above 2 in the 1950s, it has now fallen to levels close to only 1.5.

1.4 Main Risk Factors

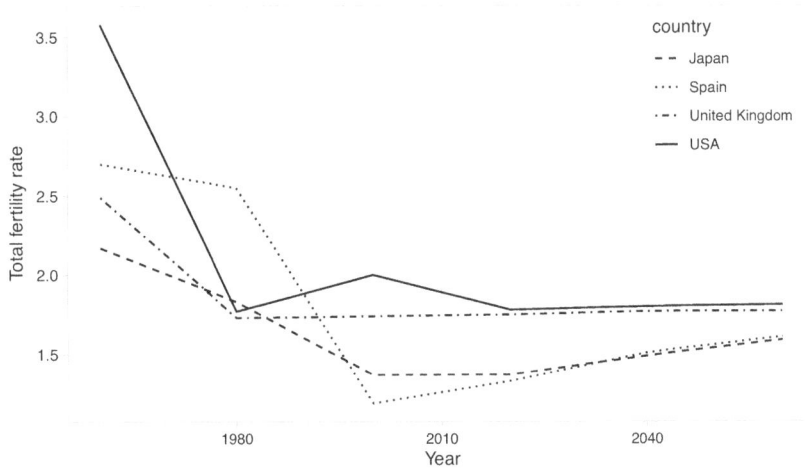

Fig. 1.3 Evolution of total fertility rate (historical and projected) from 1960 to 2060 (data source: OECD, Pensions at a glance 2021)

This increase of longevity, combined with a decrease of fertility, generates important changes in the structure of populations. Figure 1.4 compares for various countries the shape of the age pyramid in 1950 and in 2020. The aging of all of these populations between 1950 and 2020 is more than clear! We are moving from a pyramid shape to a cylinder shape with important cohort sizes above common retirement age.

As a consequence of this massive aging, there is a significant contrast between the evolution of the active population and the retired population, as suggested by Fig. 1.5. The "nearly vertical" shape of evolution in the four selected countries indicates that while the number of actives has remained nearly constant during the period 1950–2020, the number of retirees has increased significantly.

1.4.2 Economic Factors

While pure demographic figures are key elements to understand and describe quantitative issues of pensions, the job market also plays a central role. Apart from a legal retirement age existing in every country, the reality shows that many senior people leave the job market earlier. Figure 1.6 gives, for instance, the employment rates for different age groups close to a normal retirement age.

This suggests that important differences may occur between a legal retirement age and an effective exit age, as presented in Fig. 1.7.

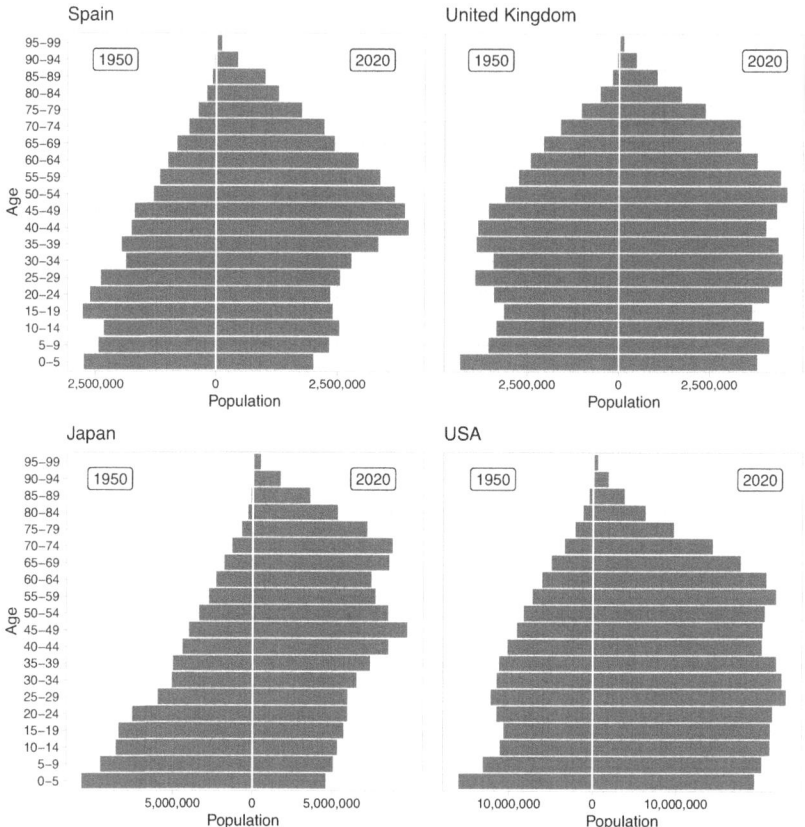

Fig. 1.4 Comparison of age pyramids in 1950 and in 2020 (data source: Human Mortality Database)

1.4.3 Financial Factors

Pensions benefits and contributions are also clearly largely dependent on the evolution of financial markets and inflation. In our modern economies, we know that financial indicators such as inflation rates or interest rates can fluctuate widely, as illustrated by Figs. 1.8 and 1.9.

1.4 Main Risk Factors

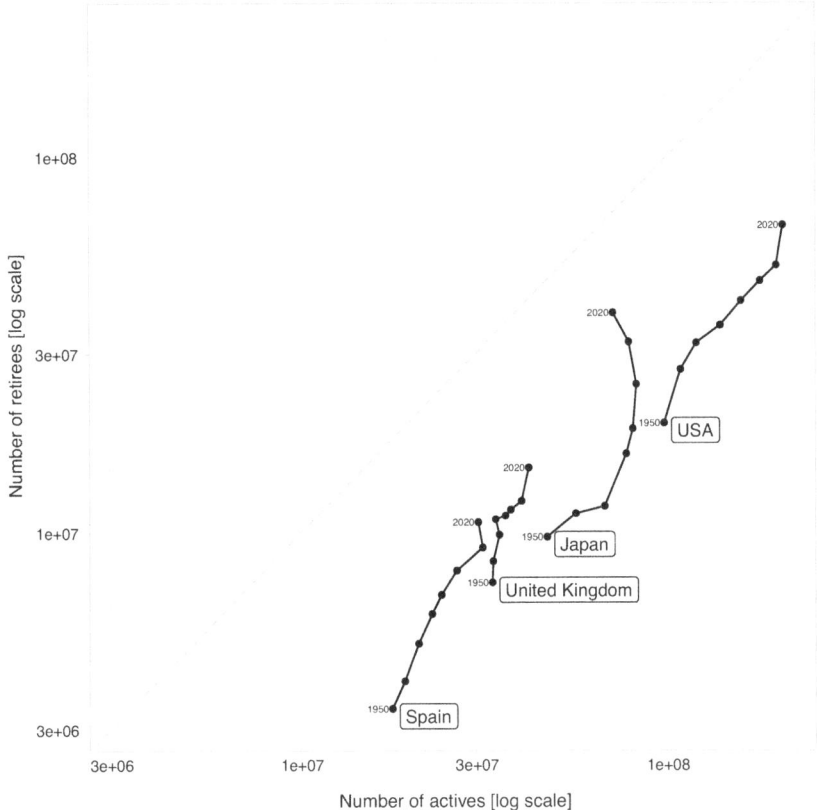

Fig. 1.5 Evolution of active and retirees populations from 1950 to 2020 (data source: Human Mortality Database)

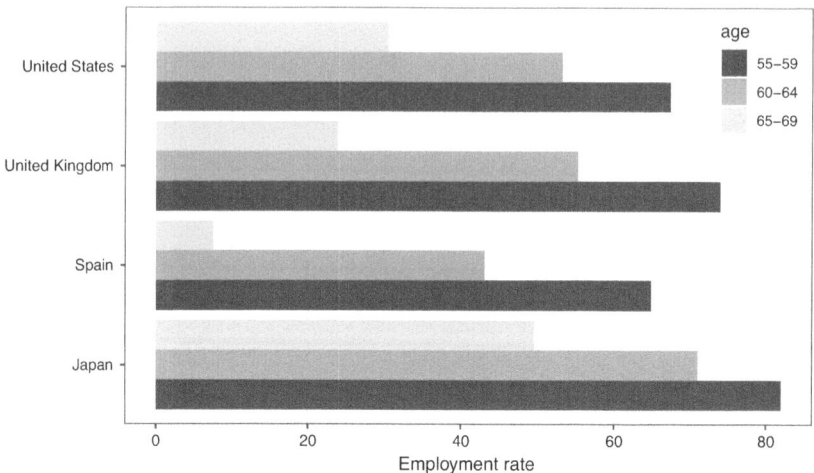

Fig. 1.6 Employment rates among seniors in 2020 (data source: OECD, Pensions at a glance 2021)

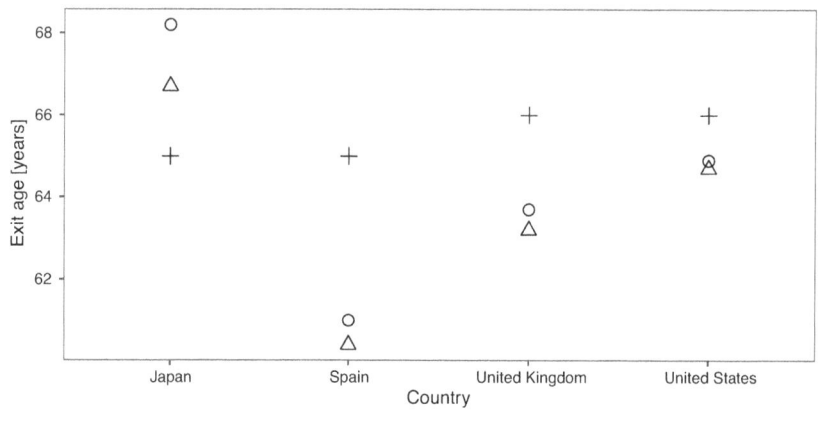

Fig. 1.7 Comparison of the effective exit ages (for men and women) and the "normal" exit age (as defined in the law) in 2020 (data source: OECD, Pensions at a glance 2021)

1.4 Main Risk Factors

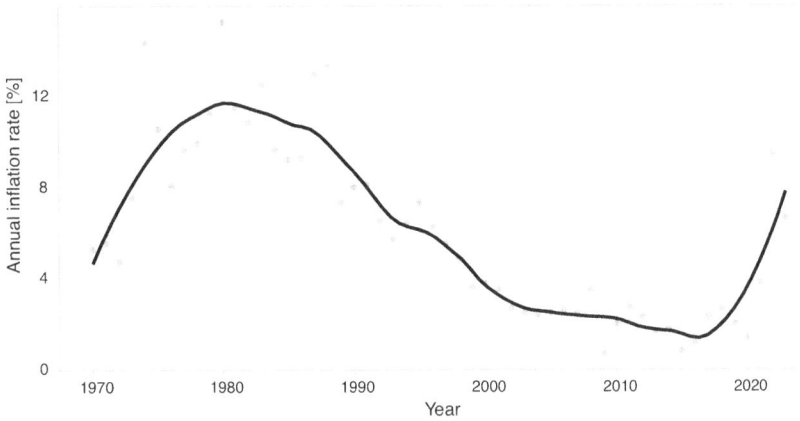

Fig. 1.8 Evolution of inflation for OECD countries 1970 to 2023 (data source: OECD)

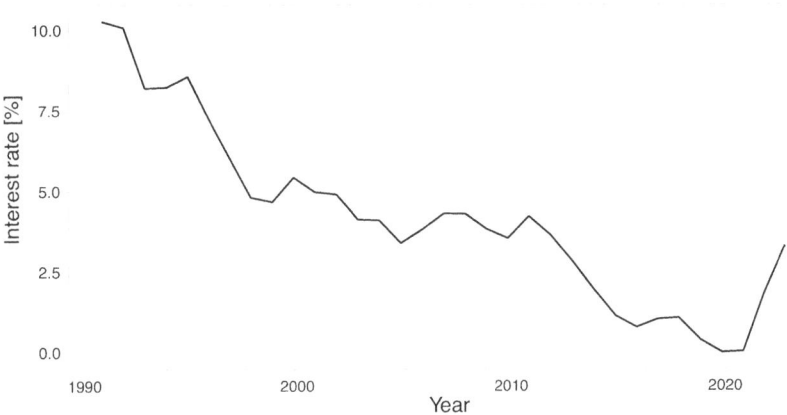

Fig. 1.9 Long term interest rate (10-year government bonds) in the Euro Zone from 1991 to 2023 (data source: OECD)

Part I
General Pension Theory

The objective of this first part is to present the foundations of the actuarial theory of pension schemes, whether they are social security schemes or supplementary pension schemes. After describing the main principles governing the management of a collective pension system and the different choices available to the manager in the effective implementation of a plan, we present some basic demographic tools useful for the valuation of pension plans. An initial classification of the major financing methods is then presented, based on the notion of actuarial equilibrium. In particular, the contrasting characteristics of Pay-As-You-Go and funded schemes are highlighted. An initial diagnosis of the choice between these two philosophies is made, based on the famous social security paradox.

Chapter 2
Pension Schemes

Abstract This second chapter introduces the concept of pension scheme by first placing it in a historical perspective. We then analyze it in terms of the various degrees of freedom available to the sponsor of a pension scheme, both in terms of the definition of the benefits granted to beneficiaries and the funding modalities.

2.1 Pensions and Pension Schemes

Pensions are based on a life cycle philosophy. The concept of pension is generated by the common idea that at a certain age, called the retirement or pension age (for instance 65), you don't have to work any more (or find a job).

In order to survive after that age, you of course need income. Basically, a pension is therefore a regular income (annual, monthly, ...) to be paid to the retiree after a given age and as long as he/she will survive. Actuarially, we will call such a benefit a lifetime annuity. Different variants or options can exist around this general concept. For instance, pension income can continue to be paid after death, partially or totally to a surviving beneficiary (reversible pension). Pensions can also sometimes be paid in the form of a unique lump sum at retirement. An important economical and philosophical question is then how to organize these pension payments optimally in a given environment.

One possibility to take care of "your old days" (... if any ...) is purely individual: you save during your active lifetime in order to have enough resources at pension age; another individual arrangement is based on family solidarity. Until the 1850s, Western societies, strongly oriented towards crafts and agriculture, organized the pension issue in this way. The doctrine of individual responsibility, assisted at a family level if necessary, predominated. The individual works as long as possible and does not reach a very old age anyway, considering the general low life expectancy at that time. Savings opportunities are limited; in a general context of poverty, almost all working income is needed to feed oneself and meet the most basic needs. The 1850–1900 period, with the development of the industrialization of our societies, saw for the first time a real collective awareness, leading to the concept of a collective pension scheme. A pension scheme can be defined as a systematic mechanism, organized on a legal or regulatory basis by a collective entity called the sponsor, to provide benefits to older people after retirement age. Individuals belonging to such a scheme will be called affiliates. In general, these schemes are contributory in the sense that for each affiliate the payment of benefits after retirement is linked (in a way to be defined) to the fact that contributions have been paid for that affiliate during his/her active career. These contributions can be paid directly by the affiliate and/or by a collective entity such as an employer. This general definition does not mean that the pension benefits must be the exact financial counterpart of the contributions paid during the active phase but a link must be clearly established between the acquisition of pension rights and the payment of contributions. The majority of pension schemes are contributory; assistance programs, giving to the poorest a minimum level of pension independently of any contribution paid before and generally financed by taxes, are examples of non-contributory schemes.

One of the first pension schemes in the modern sense was the Bismarck system created in Germany in 1889 and financed by worker and employer contributions. The twentieth century then saw a continuous development of pension systems, each country having a different history and therefore a different culture regarding this field.

Among other things, two phenomena can explain the importance that organized pension systems have taken on in modern societies as opposed to "micro-solidarity" solutions:

- the atomization of the family cell: family cells often with three generations (grandparents/parents/children) have been replaced by bi-generational cells (parents/children),
- the considerable increase in life expectancy after retirement, itself caused on the one hand by the increase in life expectancy and on the other hand by the cessation of professional activity earlier and earlier. In a simplistic way, we could say that we are retiring earlier and earlier and dying later and later.

The tremendous development of collective retirement schemes has led to the formalization of the concept of the three pillars. According to this theory, an advanced pension scheme must be based on three components, which are complementary and very different in nature:

- a first pillar: Social Security sponsored by the State, organized at the general level of a country and enabling a first basic level of pension to be granted,
- a second pillar: occupational schemes organized within a company or sector of activity, granting each of the members of these schemes a supplement to Social Security on a collective basis,
- a third pillar: individual savings organized at the free choice of each individual.

Each of these pillars has its own operating modes, the more or less collective nature of the pillar fundamentally influencing its organizational mode. The first and second pillars will each be referred to as "pension scheme" and will be the subject of this book.

2.1 Pensions and Pension Schemes

The first and second pillars may have common properties in terms of the definitions of benefits or financing structure, but are by essence quite different, mainly in two aspects:

- the size of the population involved in the scheme: a social security system is by definition characterized by a large number of affiliates (for instance all the employees of a given country); the size of an occupational pension regime can vary a lot depending on the size of the sponsor (which can be a small company or a big multinational structure),
- the time horizon of the scheme: the sponsor of a social security system (often a State or a similar public institution) can be considered as eternal; therefore, we can consider that a first pillar scheme will never stop (open risk community). On the contrary, the sponsor of a second pillar (which is in general a private company) can disappear at any time, whatever its size.

These two important differences will have major consequences when looking at the actuarial analysis of these systems. In particular, we can generate in a first pillar an intergenerational solidarity mechanism: for instance, active workers could be ready to pay the pensions of the present retirees because they count on the future generations to finance their own pension. Such arrangements are dangerous and in general not allowed in a second pillar, simply because future generations in a given company could not exist at all! Table 2.1 summarizes this comparison between Social Security pension schemes (first pillar) and occupational pension schemes (second pillar).

The importance of the first and second pillars respectively can vary greatly from one country to another, depending on historical developments. Historically, we can distinguish two basic models that have influenced the development of regimes, each responding to a certain philosophy of the role of Social Security: the Bismarckian model and the Beveridgean model.

Table 2.1 Comparison between first and second pillars

First pillar (Social Security)	Second pillar (Occupational schemes)
Organized within a country and by definition involving a large population.	Organized within a company or a sector of activity or a profession and which can concern both large numbers of employees and very small populations.
Assumed to be endless: the hypothesis of the regime's termination is not to be considered; a State claims by nature to be eternal and develops systems designed to last.	System that can stop: the hypothesis that the system will stop at any time must always be taken into account, if only by the disappearance of the company.
Open risk community: projections can be made over an infinite time horizon.	Closed risk community: projections are always made over a finite time horizon.
Possibility of transfer of charges between generations: the endless assumption of the system makes it possible to consider phenomena of solidarity between generations.	Self-financing principle: the lack of certainty as to the continuation of the scheme requires a generational balance.

2.1.1 The Bismarckian Model

In this model, the pension benefits granted by Social Security are supposed to be the counterpart given by the community, for the creation of wealth to which it is indebted to the worker. This is an insurance principle, and there is logically a significant proportionality between the income from employment and the pension benefits granted. The retirement received by the individual is the reward for working and contributing throughout his or her working life.

2.1.2 The Beveridgean Model

The Beveridgean philosophy for social security is based on the vision that the first pillar has essentially to be a safety net against elderly poverty. The idea was first introduced in the United Kingdom during the second World War by the minister Beveridge in an important context of poverty. The purpose of such a social security scheme is therefore not to provide pension benefits linked to career salaries, as in the Bismarckian model, but to guarantee a minimum level of benefits for everyone, irrespectively of the intensity of work before retirement. These systems can be considered to be based on an assistance principle rather than a social insurance principle. In general, the pension benefits are much more uniform.

2.1.3 Models Around the World

Most modern systems simultaneously take these two models into account through pensions calculated in proportion to wages but also taking into account minimum amounts. The mixing in different proportions of the two models illustrates the two natural basic functions of a pension scheme: to deliver pension benefits in line with career salaries and to protect against elderly poverty. It is however easy to distinguish in each country a predominant influence of one or the other model. Modern social security pension schemes can be classified into four categories:

1. Pure occupational systems: the organization of pension schemes is based exclusively on the labor market and the different occupational categories. Retirement is directly linked to activity wages. The system's sole objective is to maintain part of the activity income. Contributions are paid by members and their employers (example: Germany).
2. Bismarckian systems with safety net: the organization is the same, but there are also minimum pension rules for the poorest people, which are not necessarily linked to a professional activity (example: France, Belgium, Italy, Spain).
3. Universal systems with professional organization: a basic Beveridgean scheme covering the whole population is provided for through a first pillar financed

by contributions. A second, almost generalized, pillar is organized at the professional level (e.g. United Kingdom, Netherlands).
4. Purely universal systems: a single basic Social Security system exists for all citizens and is financed by taxation; a uniform and state-administered second pillar covers the working population (e.g. Sweden, Norway).

It is easy to see that, depending on the underlying philosophy, the respective positions of the first and second pillars can vary greatly from one country to another.

2.2 Design of the Pension Scheme

When a pension scheme is set up, whether it is a first or second pillar scheme, the sponsor has to answer many questions about its structure:

- What benefits philosophy should be considered?
- How is the size of these benefits defined?
- How are these promises funded?
- What legal vehicle should be used to establish the system?

These are some of the questions we propose to address in the rest of this chapter.

The first task in setting up a pension scheme is to define the philosophy supporting the pension benefits to which scheme members will be entitled. Depending on whether one wishes to focus on the replacement income role or on the savings accumulation role of the scheme, two systems can be distinguished:

- Defined Benefit schemes,
- Defined Contribution schemes.

2.2.1 *Defined Benefit Schemes (DB)*

These schemes, with a replacement income approach, explicitly define the benefits to which future retirees will be entitled. These benefits can be flat-rate (e.g. a fixed monthly amount under a minimum old-age income scheme for all). They are usually defined according to different parameters such as length of membership and earned salaries.

The following formula can be considered as an example:

Retirement pension = 30% of the average salary over the last five years of

employment.

With benefits defined in this way, an important question is how to finance them: the funding of Defined Benefit schemes is one of the most important aspects of actuarial pension theory. It will be the subject of long developments in the rest of this book.

The advantage of such schemes is clearly transparency towards members, who are fully aware of the level of benefits they will receive. Conversely, the cost is unknown ex ante... and may have surprises in store.

It should be noted that some DB pension schemes are "non-contributive", in the sense that the affiliates do not have to pay contributions in order to receive benefits when retired. In the rest of this book, we will concentrate mainly on contributive schemes where actuarial methods are used.

2.2.2 Defined Contribution Schemes (DC)

These schemes, which are based on a saving approach, explicitly define the contributions that will be paid to the scheme during the working period. The services obtained are derived from it according to defined rules. It is therefore the level of funding that is defined ex ante, with benefits having to be adjusted to ensure the balance of the system.

The following formula can be considered as an example:

$$\text{Retirement contribution} = 10\% \text{ of salary.}$$

The advantage of such schemes is of course the control of cost, which is known in advance. Their drawback is a relative uncertainty for affiliates about the actual level of benefits granted by the scheme at retirement age.

Another way to define pure DC schemes is based on the following legal condition: in a DC plan, the liability of the sponsor stops at each payment of contributions. In a DB plan, the liability of the sponsor continues until the payment of the benefits.

2.2.3 Hybrid Schemes

Besides these two design types, we also find hybrid schemes, which can be considered as compromises between DB and DC schemes. Hybrid schemes can exist both in the first and in the second pillar. Here are some examples of such architectures:

- DC with guarantees: in a pure DC scheme, the liability of the sponsor stops, for a given year of work, at the payment of the contribution of that year; no additional corrections will be needed afterwards. A variant of this method is to add a minimum guaranteed return on each contribution at retirement age.

- Cash Balance: The system is similar to a DC scheme but on a notional level. Notional contributions are defined as in a pure DC system (for instance yearly contribution of 5% of the salary); these contributions are then capitalized till retirement age, not using a real rate of return corresponding to financial assets, but using a notional fixed interest rate or a notional index.
- Social security hybrid schemes: in these systems, both contributions and benefits may be adjusted with respect to their initial conditions, in order to take into account inter-generational fairness. These techniques will be presented in Chap. 7.

2.3 Benefits of the Pension Scheme

Once the fundamental choice has been made between a Defined Benefit scheme or a Defined Contribution scheme, we must define the parameters of the scheme.

2.3.1 Defined Benefit Schemes

Typically, these schemes fix a retirement pension, which can be defined as a periodic income payable from the age of retirement until the death of the beneficiary. Often schemes also provide for a reversion of this pension to the surviving spouse: in the event of the predecessor's death, a percentage of the pension (e.g. 75%) continues to be paid to the surviving spouse until death. On the other hand, the pension can be indexed after retirement, either on a flat-rate basis (for example, a pension increasing annually by 2%) or by following the evolution of a specific index. The two main parameters to define in a DB plan are the retirement age (denoted by x_r) and the pension amount (denoted by R), which is a lifetime annuity to be paid from retirement age till death (regular income, for instance yearly).

The retirement pension is a function of several variables, generally:

- the salary of one or more years of career,
- the length of the career,
- ceilings to cap or slice salaries,
- benefits obtained from other schemes.

A fairly general retirement pension formula can be expressed as follows:

$$R = \alpha f(N) g(S) - M,$$

where

- R is the annual pension at retirement age,
- N is the career duration taken into account in the scheme,

- S is the career salary vector,
- $f(N)$ is a function of the career duration,
- $g(S)$ is a function of the salary,
- α is a constant,
- M is a possible amount to be deducted.

The three main choices to be made are therefore:

1. the definition of the reference salary (function g),
2. the definition of career length (function f),
3. the possible subtractive elements (M term).

2.3.1.1 Definition of the Reference Salary

Denoting by S_j ($j = x_0, \ldots, x_r - 1$) the vector of salaries between the age of affiliation and the retirement age, the salaries to be taken into account in the retirement formula are generally determined by the following two operations.

Capping and/or Slicing

Salaries can be capped:

$$\hat{S}_j = \min(S_j; P_j),$$

where P_j is the cap for the year j (same amount for all).

They can also be sliced:

$$\hat{S}_j = k_1 \min(S_j; P_j) + k_2 \max(S_j - P_j; 0),$$

where P_j is the cap for the year j and k_1, k_2 are two coefficients defined in the scheme.

For example:

Reference salary = 100% annual salary up to 25,000 + 75% of the balance.

Number of Salaries Taken into Account

The following systems are commonly found:

- career average: the reference salary is the arithmetical average of the potentially capped salaries over the entire career:

$$g(S) = \frac{1}{x_r - x_0} \sum_{j=x_0}^{x_r - 1} \hat{S}_j,$$

2.3 Benefits of the Pension Scheme

- indexed career average: historical salaries are indexed at the date of retirement:

$$g(S) = \frac{1}{x_r - x_0} \sum_{j=x_0}^{x_r-1} \hat{S}_j g_j,$$

where g_j = indexation factor between age j and retirement age x_r.
- final salary: only the last active salary is taken into account:

$$g(S) = \hat{S}_{x_r-1},$$

- average of the last few years: we take into account the last n salaries:

$$g(S) = \frac{1}{n} \sum_{j=1}^{n} \hat{S}_{x_r-j}.$$

Using career average or final salary fundamentally changes the level of benefits, as illustrated by Example 2.1.

Example 2.1 Consider a scheme providing a retirement pension equal to 75% of the average career salary. What does this scheme represent, expressed as a percentage of the final salary? We use the following assumptions:

- initial age: $x_0 = 20$,
- retirement age: $x_r = 60$,
- initial unit salary $S_{20} = 1$, increasing in geometric progression by a factor of $(1+g)$.

The pension is given by

$$R = 75\% \frac{1}{40} \sum_{k=0}^{39} (1+g)^k$$

$$= 75\% \frac{1}{40} \frac{(1+g)^{40} - 1}{g}.$$

Expressed as a percentage of final salary:

$$\gamma = \frac{R}{S_{59}} = 75\% \frac{1}{40} \frac{(1+g)^{40} - 1}{g(1+g)^{39}}.$$

Table 2.2 gives the value of γ as a function of the wage growth rate. Thus, for a salary increase rate of 5%, a scheme based on 75% of the career average only delivers 34% of the last salary.

Table 2.2 Value of α as a function of the wage growth rate

g	γ
2%	52%
5%	34%
10%	20%

Table 2.3 Salary evolution

Age	Salary
20–29	1
30–44	1.2
45–54	1.5
55–58	1.8
59	1.9

Under the classical assumption of increasing salaries during the career, it is clear that a final salary scheme looks better for the affiliate. On the other hand, a scheme based on final salary, while better protecting the affiliate, may generate surprises for the sponsor in terms of cost, as shown in Example 2.2.

Example 2.2 We use the following assumptions:

- career between ages 20 and 60,
- initial unit salary $S_{20} = 1$, increasing according to Table 2.3,
- the scheme provides a pension equal to 50% of final salary,
- we assume that funding is provided during the career by constant payments based on the salary known at the time of calculation (the last known salary is used at any time as an estimator of the final salary),
- the discount rate is 4%,
- we assume there is no mortality before retirement,
- the annuity price[1] at age 60 (allowing at retirement age the conversion of accumulated savings into periodic income until death) is assumed to be equal to 10: $a_{60} = 10$.

We are interested in the evolution of the contribution rate of the scheme for the affiliate, i.e. the ratio between the contribution to be paid and the salary. The initial contribution is given by (see technical appendix):

$$p_{20} = \frac{0.5 \cdot v^{40}}{\ddot{a}_{\overline{40|}}} 10 = 0.051,$$

---

[1]The annuity price is a conversion coefficient of a regular lifetime pension into a lump sum. See the technical appendix for details.

(continued)

2.3 Benefits of the Pension Scheme

Example 2.2 (continued)
where $v = 1/1.04$ and

$$\ddot{a}_{\overline{40|}} = \frac{1 - v^{40}}{0.04} \cdot 1.04.$$

The initial contribution rate (ratio between the contribution and the salary) is therefore 5.1%.

At age 30, the salary increase requires that additional contributions must be made between this age and retirement age:

$$p_{30} = p_{20} + \frac{0.5 \cdot 0.2 v^{30}}{\ddot{a}_{\overline{30|}}} \cdot 10 = 0.051 + 0.017 = 0.068.$$

The contribution rate becomes:

$$\pi_{30} = \frac{0.068}{1.2} = 0.056.$$

Similarly, at subsequent ages, we obtain the contribution rates given in Table 2.4. Salary increases at the end of a career therefore cause an explosion of the cost in this kind of scheme.

Finally, it should be noted that some DB pension schemes do not take salaries into account at all. This is the case with schemes of the "flat amount per year" type, where a lump sum is granted per year of membership.

2.3.1.2 Definition of Career Length

Generally, benefits are related to career length, using fractions of the type:

$$\frac{N}{N_{\text{tot}}},$$

where N = career length taken into account for the affiliate and N_{tot} = maximum career length taken into account in the pension scheme.

Table 2.4 Contribution rates obtained in Example 2.2

Age	Contribution rate
20–29	5.1%
30–44	5.6%
45–54	9.3%
55–58	22.6%
59	46.6%

This factor is based on a natural "proportionality" (proportional aspect of benefits to the duration of service), even if, from an actuarial point of view, a higher penalty could be justified. Example 2.3 illustrates this phenomenon.

Example 2.3 We use the same assumptions as in Example 2.2. The scheme provides a retirement pension at retirement age of 60 years equal to

$$R = \frac{N}{40} 50\% S,$$

where N = career length between entry into service and retirement age, and S = last career salary.

We consider two affiliates entering into the scheme at $t = 0$, one 20 years old, the other 40 years old. We compare the contribution rates to the scheme, for a constant unit salary:

1. for the 20-year-old affiliate:

$$\pi_{20} = \frac{0.5 v^{40}}{\ddot{a}_{\overline{40|}}} \cdot 10 = 0.051,$$

2. for the 40-year-old affiliate:

$$\pi_{40} = \frac{20}{40} \cdot \frac{0.5 v^{20}}{\ddot{a}_{\overline{20|}}} \cdot 10 = 0.081.$$

Despite the 50% correction in benefits, the contribution is nearly 60% higher for the 40-year-old member.

As far as the definition of the length of career N is concerned, there are different possibilities. The problem typically arises when a scheme is put in place when employees already have years of service in the company. In this case, we can:

- only count years of service after joining the scheme:

$$N = N_2 = \text{age at retirement} - \text{age at membership},$$

- also count the years between the start of service and the age of affiliation:

$$N = \text{age at retirement} - \text{age at entry into service} = N_1 + N_2,$$

where N_1 = age at membership − age at entry into service. This is referred to as recognition of initial past service (or back service),

2.3 Benefits of the Pension Scheme

- revalue only a portion of these past services:

$$N = N_2 + \min(N_1; N_{\max}),$$

where $N_{\max}$ is a fixed number of years for all affiliates (e.g. 10 years).

Depending on the age at entry into service and age at membership, funding initial past service can represent a very significant cost, as illustrated in Example 2.4.

Example 2.4 We use the same assumptions as in Example 2.3. We assume that all members entered service at 20 years, and that the scheme recognizes past service. For an affiliate aged 40 at the initiation of the scheme, we have:

- past service: $N_1 = 20$,
- future service: $N_2 = 20$.

The contribution rate for future service is:

$$\pi_{40}(N_2) = \frac{20}{40} \cdot 10 \cdot \frac{0.5 v^{20}}{\ddot{a}_{\overline{20}|}} = 0.081.$$

The contribution rate for past service is:

$$\pi_{40}(N_1) = \frac{20}{40} \cdot 10 \cdot \frac{0.5 v^{20}}{\ddot{a}_{\overline{20}|}} = 0.081.$$

The total contribution rate is:

$$\pi_{40} = \pi_{40}(N_1) + \pi_{40}(N_2) = 0.162.$$

Table 2.5 shows the contribution rates by age at the time the scheme was implemented. Column (1) is given by

$$\pi_x(N_1) = \frac{x - 20}{40} \cdot \frac{0.5 v^{60-x}}{\ddot{a}_{\overline{60-x}|}} \cdot a_{60}.$$

Column (2) is given by

$$\pi_x(N_2) = \frac{60 - x}{40} \cdot \frac{0.5 v^{60-x}}{\ddot{a}_{\overline{60-x}|}} \cdot a_{60}$$

with $a_{60} = 10$ and $v = 1/1.04$. Column (1) of Table 2.5 shows the exponential growth in the past service contribution rate.

Table 2.5 Contribution rates obtained in Example 2.4

Age x at initiation of the scheme	Past service contribution rate (1)	Future services contribution rate (2)	Total contribution rate (1) + (2)
20	0	5.1%	5.1%
30	2.1%	6.4%	8.6%
40	8.1%	8.1%	16.2%
50	30%	10%	40%
55	77.7%	11.1%	88.8%

2.3.1.3 Subtractive Elements

As pension schemes can overlap (e.g. a second pillar scheme in addition to a basic first pillar scheme), it is not uncommon for some formulas to take into account the benefits already guaranteed by other schemes.

Two techniques are traditionally used in this respect: the "offset" formula and the "step rate" formula.

Offset Formula

Benefits generated by another plan are explicitly deducted in the pension formula. Such a plan is given in Example 2.5.

Example 2.5 Retirement pension given by:

$$R = \frac{N}{40} 70\% S - R_{SS},$$

where N = career length, S = last salary = $S_{x_r - 1}$, and R_{SS} = pension granted by the general Social Security system.

The objective of this scheme is therefore to provide an overall retirement level corresponding to 70% of the last salary, including the first pillar pension.

Step Rate Formula

Generally, Social Security schemes only take salaries into account in benefit formulas up to a certain cap. Therefore, the salary bracket above this cap does not give entitlement to any social security pension. In this context, it seems natural, when setting up a complementary second pillar scheme, to grant greater benefits on the salary bracket above the cap.

Example 2.6 shows how such a construction can be justified.

2.3 Benefits of the Pension Scheme

Example 2.6 As in Example 2.5, we want to grant an overall pension level corresponding to 70% of the last salary. It is assumed that the basic scheme provides, up to a certain compensation cap denoted P, a retirement pension corresponding for a 40-year career at 50% of the last salary.

So in this case, we have:

$$R_{SS} = \frac{N}{40} 50\% \cdot \min\left(S_{x_r-1}, P\right).$$

The objective of the second pillar scheme can then be written:

$$R = \frac{N}{40} 70\% S_{x_r-1} - \frac{N}{40} 50\% \min\left(S_{x_r-1}, P\right)$$

$$= \frac{N}{40} \left[20\% \min\left(S_{x_r-1}, P\right) + 70\% \max\left(S_{x_r-1} - P, 0\right)\right].$$

Compared to the offset formula, this formula has a double advantage:

- a greater simplicity of calculation and communication to affiliates: it depends only on a general parameter of the basic scheme, the cap. It is no longer necessary to calculate for each affiliate the precise first pillar formula,
- a less direct dependence on the basic scheme, especially if the parameters of the latter are modified.

2.3.1.4 Annuity or Lump Sum

The examples presented above express the benefits in the form of an annuity (periodic income payable from retirement age). In terms of guaranteeing replacement income, this is obviously the most natural form of pension scheme benefit. However, there are alternatives in the form of a lump sum benefit.

This choice between annuity and lump sum can be made at different stages.

In the Definition of the Retirement Formula

Rather than guaranteeing a regular pension payable from retirement age, a single lump sum, payable once at retirement age, is defined.

Example 2.7 Guarantee at retirement age of a lump sum corresponding to three times the average salary of the last five years of activity, for a career of 40 years:

$$L = \frac{N}{40} 3 \left(\frac{1}{5} \sum_{i=1}^{5} S_{x_r - i} \right).$$

In the Way of Payment of the Benefits

Some schemes, even if they are initially formulated in the form of an annuity (or lump sum), may offer the member the option of receiving benefits rather in the form of a lump sum (or pension).

The annuity conversion factor is called the annuity price.

Example 2.8 The lump sum-based scheme in Example 2.7 can generate at retirement age a pension equal to

$$R = \frac{1}{a_{x_r}} L.$$

The pension can be a single life annuity (that of the retiree). It can also be reversible to a certain percentage for the benefit of the surviving spouse. It can also be constant or increasing.

Assuming, for example, a reversion rate of 2/3 for the benefit of the spouse (of the same age by assumption), the annuity price to be used for the conversion of the lump sum into an annuity becomes:

$$a_{x_r} + \frac{2}{3} \left(a_{x_r} - a_{x_r x_r} \right),$$

where $a_{x_r x_r}$ is the price of an annuity on two heads.

2.3.1.5 Some DB Standard Formulas

By combining the elements described above, various benefit formulas can be generated. We give below some typical schemes as examples.

2.3 Benefits of the Pension Scheme

Scheme 1: Fixed Amount Formula

$$R = \frac{N}{45} F,$$

where R = annual retirement pension; N = career duration since joining the scheme, limited to 45 years; F = constant amount equal, for example, to 5000 euros.

Scheme 2: Final Salary Formula

$$R = \frac{N}{40} 50\% S_{x_r - 1},$$

where R = annual retirement pension, N = length of career since entering service (capped to 40 years), and $S_{x_r - 1}$ = last annual active salary of the career before retirement age x_r. Writing the formula in this form suggests the overall objective of the scheme for a full 40-year career: to get half of the final salary.

Alternatively, the objective can be written as

$$R = N \left(1.25\% S_{x_r - 1}\right),$$

suggesting that each year of service generates a pension share equal to 1.25% of the last salary.

Scheme 3: Sliced Final Salary Formula

$$R = \frac{N}{40} \left(20\% \min \left(S_{x_r - 1}; P\right) + 75\% \max \left(S_{x_r - 1} - P; 0\right)\right),$$

where R = annual retirement pension, N = length of career since joining the scheme (capped to 40 years), $S_{x_r - 1}$ = last annual active salary of the career before retirement age x_r, P = salary cap, set for instance to 25,000 euros.

The objective of the scheme is to obtain a retirement pension for a full career equal to 20% of the last salary up to the cap, and 75% of the portion of the last salary exceeding this cap.

Scheme 4: Average Capped Salary for the Last 5 Years Formula

$$R = \frac{N}{45} 60\% \overline{S}(5),$$

where R = annual retirement pension, N = length of career since the date of affiliation (capped to 45 years), $\overline{S}(5)$ = average of the last five annual salaries of activity, each of them being capped to a maximum amount of 40,000 euros.

The objective of the scheme is to obtain a retirement pension for a full career equal to 60% of the average of the last five salaries in the career, each of which is capped to a maximum of 40,000 euros.

Scheme 5: Career Average Formula

In this case, we have

$$R = \frac{N}{40} 75\% \overline{S},$$

where $R=$ annual retirement pension, $N =$ length of career since age of entry into service x_0, and $\overline{S} =$ average salary.

Once again, this formula explains the scheme's ambition for a full 40-year career: to obtain 3/4 of the average salary.

Alternatively, observing that $N = x_r - x_0$, we can write

$$R = 1.875\% \sum_{x=x_0}^{x_r-1} S_x,$$

suggesting now that each year of service generates a pension share corresponding to 1.875% of that year's salary.

Scheme 6: Offset Final Salary Formula

$$R = \frac{N}{40} 80\% S_{x_r-1} - R_{SS},$$

where $R=$ annual retirement pension, $N =$ length of the career since entering service (capped to 40 years), $S_{x_r-1} =$ last annual salary of the career before retirement age x_r, and $R_{SS} =$ pension obtained from the basic Social Security scheme.

The objective of the scheme is to obtain (for a full career) 80% of the last salary, taking into account the pension benefit already granted by the general Social Security system.

Scheme 7: Sliced Lump Sum Formula

$$L = \frac{N}{45} \left(2 \min \left(S_{x_r-1}; P \right) + 5 \max \left(S_{x_r-1} - P; 0 \right) \right),$$

where $L =$ lump sum available at retirement age, $N =$ length of career since joining the scheme (capped to 45 years), $S_{x_r-1} =$ last annual salary of the career before retirement age x_r, and $P =$ salary cap, set for instance to 25,000 euros.

The objective of the scheme is to obtain at retirement age a lump sum equal to twice the last salary up to the cap, and 5 times the part of salary exceeding this cap. This lump sum available at retirement age can be converted into a pension annuity using a pension conversion factor:

$$R = \frac{L}{a_{x_r}}.$$

2.3.2 Defined Contribution Schemes

In a Defined Contribution scheme, the plan must define the level of contribution to be paid for affiliates. In practice, the following formulas are mainly used:

- contributions as a percentage of salaries: for example,

$$\text{pension contribution} = 5\% \text{ salary},$$

- contributions as a percentage of capped salaries: for example,

$$\text{pension contribution} = 7\% \min(\text{salary}; P),$$

where P is a salary cap.
- sliced contributions:

$$\text{pension contribution} = 5\% \min(\text{salary}; P) + 10\% \max(\text{salary} - P; 0).$$

- increasing contributions: the percentage of contributions to be applied to the salary increases with age. For example:

$$\text{pension contribution} = \begin{cases} 2\% \text{ salary} & (20 < x < 35), \\ 3\% \text{ salary} & (35 < x < 50), \\ 4\% \text{ salary} & (x > 50), \end{cases}$$

where x is the age of the affiliate.
- sliced and increasing contributions: a combination of the above two variants.

The expression "Defined Contribution" also include profit sharing schemes. These schemes define rules for the distribution of an annual amount (which is not linked to the payroll, but to the company's results) to all affiliates.

Finally, it should be noted that, like Defined Benefit schemes, Defined Contribution schemes may, depending on the case, provide for a payment in the form of lump sum and/or annuities.

While Defined Contribution schemes have the great merit of inducing cost transparency by essence, they can lead to questionable benefit hierarchies.

Example 2.9 illustrates the highly variable level of benefits that a Defined Contribution system can generate, depending on the member's career profile.

Table 2.6 Career profiles

Age	Career A	Career B	Career C
20–29	100	75	100
30–39	125	100	200
40–49	125	150	400
50–59	125	150	600

Example 2.9 Let us consider a DC scheme where successive contributions are invested and generate at retirement age an individual saving for each affiliate. This saving is then converted into a lifetime annuity. The parameters are defined as follows:

- contribution: $\pi = 10\%$ of salary,
- rate of return: $r = 4\%$,
- three categories of employees rated A, B and C,
- age of affiliation: $x_0 = 20$,
- retirement age: $x_r = 60$,

The savings accumulated at retirement are converted into a lifetime annuity using a conversion factor (annuity price). It is assumed that the annuity price at age 60 is 10 ($a_{60} = 10$). The career profiles of the 3 categories are shown in Table 2.6. The effect of mortality during the active period is ignored.

Denoting by S_{ij} the salary at age j for the category i, we can compute the following quantities:

- sum of career salaries:

$$\Sigma_i^{(1)} = \sum_{j=20}^{59} S_{ij},$$

- career average salary:

$$\Sigma_i^{(2)} = \frac{1}{40} \sum_{j=20}^{59} S_{ij},$$

- savings accumulated at retirement:

$$\Sigma_i^{(3)} = \sum_{j=20}^{59} \pi S_{ij} \cdot (1+r)^{60-j},$$

(continued)

2.3 Benefits of the Pension Scheme

Example 2.9 (continued)
- pension obtained at age 60:

$$\Sigma_i^{(4)} = \frac{1}{a_{60}} \Sigma_i^{(3)} = \frac{1}{10} \Sigma_i^{(3)}.$$

Table 2.7 shows some of these elements.

The pension obtained can be expressed as a percentage of the last active salary. This ratio is called the "replacement rate" and is given in Table 2.8.

It can be seen that the system is very unfavorable to dynamic career profiles and prefers flat profiles instead. Indeed, the system makes it possible to generate a pension income corresponding to more than 90% of the last salary for the first category, while it only leads to a low level of 40% for the third category that has experienced the most significant career development.

Another issue of Defined Contribution schemes is to expose affiliates to financial market risks. Using the example above and varying the pre-retirement rate of return (the annuity price at age 60 remaining fixed at 10), we obtain the replacement rates shown in Table 2.9.

Rather than considering the simplistic career profiles seen above, we can introduce a geometrically progressive salary function. We consider, for example, the salary at age j given by:

$$S_j = (1+g)^{j-20}.$$

We can then study the replacement rate as a function of the rate of return r and the rate of salary increase g, as in the following formula:

$$\frac{\sum_{i=20}^{59} 0.1(1+g)^{j-20}(1+r)^{60-j}}{a_{60}(1+g)^{39}}.$$

Table 2.10 illustrates this formula. It clearly shows the volatility of the replacement rate in Defined Contributions according to the parameters of the financial environment.

Another factor having a major impact on the level of benefits in DC is the length of the period during which contributions are paid. Table 2.11 gives the replacement rate of the scheme for a funding rate of 4%, a salary increase rate of 2%, and various contribution periods less than or equal to the 40 years considered above. This table shows that the reduction is more than proportional to the length of the career.

Table 2.7 Benefits by career category

	A	B	C
$\Sigma_i^{(1)}$	4750	4750	13,000
$\Sigma_i^{(2)}$	118.75	118.75	325
$\Sigma_i^{(4)}$	111.2	102.2	239.4

Table 2.8 Replacement rate by career category

	A	B	C
$\Sigma_i^{(4)}/S_{i59}$	91%	69%	41%

Table 2.9 Replacement rate as a function of rate of return

	A	B	C
$r = 6\%$	148%	109%	59%
$r = 5\%$	115%	87%	49%
$r = 4\%$	91%	69%	41%
$r = 3\%$	72%	56%	34%
$r = 2\%$	57%	46%	29%

Table 2.10 Replacement rate as a function of rate of return and salary growth rate

	$g = 0\%$	$g = 2\%$	$g = 4\%$	$g = 6\%$
$r = 6\%$	164%	99%	63%	42%
$r = 5\%$	127%	78%	51%	35%
$r = 4\%$	99%	62%	42%	29%
$r = 3\%$	78%	50%	34%	25%
$r = 2\%$	62%	41%	29%	21%

Table 2.11 Replacement rate and length of the period during which contribution is paid

N		%
40	62%	100%
30	42%	67%
20	25%	40%
10	11%	18%

2.4 Financing the Pension Scheme

In a Defined Benefit (contributive) scheme, once the benefits have been defined, the main actuarial question is then how to finance them, in other words, how the contributions have to be fixed in order to meet the scheme's promises. Similarly, in a Defined Contribution scheme, the benefits must be determined on the basis of the contributions.

2.4.1 Actuarial Equilibrium

The link between benefits and contributions in a contributive pension scheme is based on the following two fundamental concepts:

2.4 Financing the Pension Scheme

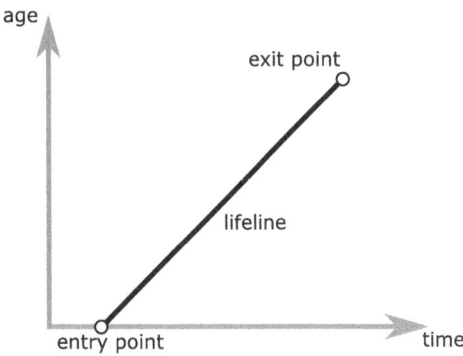

Fig. 2.1 Lifeline in the Lexis diagram

1. the risk community,
2. the actuarial equivalence relationship.

The risk community is a subset of the population which is aggregated for the financing of their retirement, becoming therefore interdependent.

The actuarial equivalence relationship then expresses, for each risk community, the equilibrium between benefits and contributions:

$$\text{present value of contributions for the community} = \text{present value of benefits for the community}.$$

Unlike traditional actuarial principles used in life insurance, there is no requirement here for a contribution/benefit equilibrium at the level of each individual (or more precisely at the level of a cohort of individuals of the same age); equivalence is considered here at the level of each community to be defined and can therefore be very collective.

The choice of the type of risk community will lead to different financing methods. Choosing a financing method therefore means choosing communities interdependent in financing their retirement.

The Lexis diagram is a useful tool to easily visualize risk communities. It is a two-dimensional graph, the abscissa representing current time, the ordinate age. A lifeline in this graph represents the evolution of an individual over time (Fig. 2.1).

2.4.2 Fully Funded or Pay-As-You-Go Scheme?

Two extreme classes of methods generated by very different risk communities can be distinguished.

2.4.2.1 Fully Funded Methods

The contribution/benefit equilibrium is achieved at the level of each individual's lifeline. We can therefore say that everyone finances their own retirement. If there is solidarity, it is at the level of the cohort of all individuals born in the same year. However, there is no solidarity between generations.

A risk community can be represented by a diagonal in the Lexis diagram (Fig. 2.2).

2.4.2.2 Pay-As-You-Go Methods

The contribution/benefit equilibrium is achieved at all times, by comparing the contributions paid by the working population and the benefits to be paid to retirees.

A risk community is therefore made up of all contributors and retirees living at the same time and can be represented by a vertical in the Lexis diagram (Fig. 2.3).

Fully-funding and Pay-As-You-Go present very different characteristics. Here are some elements among others, showing their very different behavior:

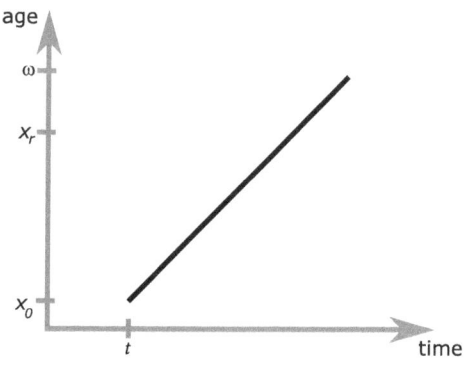

Fig. 2.2 Lexis diagram of one fully funded risk community

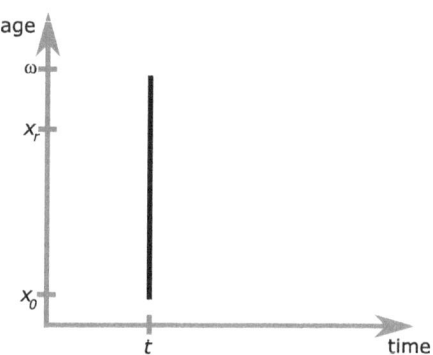

Fig. 2.3 Lexis diagram of one Pay-As-You-Go risk community

2.4 Financing the Pension Scheme

- inflation risk: in PAYG, pensions are financed by contributions of active workers; therefore, it is easy to increase the pension benefits simultaneously with the salaries. In fully funded systems, pension benefits are financed through past provisions and cannot be adapted so easily to inflation rate,
- demographic risk: in PAYG, the demographic ratio between the number of retirees and the number of contributors is crucial. In a fully-funded philosophy each generation is autonomous,
- main message: PAYG is based on an intergenerational solidarity principle (solidarity at any time between the present retirees and the present active workers). Fully funding is more oriented to an individual equity; each generation finances its own pension,
- time horizon: PAYG is naturally oriented to an infinite time horizon; you accept to pay today contributions to finance the pensions of other people because you know that at your retirement other people will pay for you. Fully funding can stop at any time; each generation will then receive the counterpart of their own contributions,
- retroactivity: In case of creation of a new pension scheme, PAYG permits to pay directly full benefits to the first generation of retirees, even if these affiliates have never paid any contributions before. This is not possible in fully funding. Those who have never contributed in the past will normally not receive any pension benefit,
- provisioning: pure PAYG does not theoretically generate any provision, being based on an immediate transfer of money from the active workers to the retirees. Fully funding implies important savings from the active workers in order to constitute their own pension.

Table 2.12 summarizes these elements.

It is quite common in the pension debate to passionately oppose these two systems, which, by essence, have both extreme and contrasting features. However, there is a whole range of intermediate actuarial methods between these two extremes, which can be grouped into three families.

Table 2.12 Comparison between Pay-As-You-Go and fully funded schemes

Pay-As-You-Go scheme	Fully funded scheme
No dependence on inflation	High dependence on inflation
High dependence on demographic change	Weak dependence on demographic change
Intergenerational solidarity	Equity at the individual level
Underlying assumption of system permanence	Non necessary permanence
Easy retroactivity in case of scheme creation	Difficult retroactivity in case of scheme creation
No provisioning	Significant provisions

2.4.2.3 Scaled Premium Pay-As-You-Go Methods

One risk community is made up of all the active and retired people present, no longer just in an instant, but over a period of T years (Fig. 2.4).

2.4.2.4 Terminal Funding Pay-As-You-Go Methods

One risk community is made up, on the one hand, of all the active workers present in an instant and, on the other hand, of the generation retiring at that moment (Fig. 2.5).

2.4.2.5 Collective Funded Schemes

The risk community is made up of the lifelines of several cohorts which are aggregated for the financing of their retirement (Fig. 2.6).

All these actuarial methods will be developed in Chap. 4.

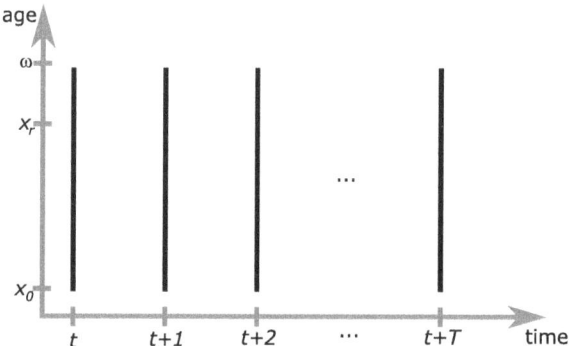

Fig. 2.4 Lexis diagram of one Scaled Premium Pay-As-You-Go risk community

Fig. 2.5 Lexis diagram of one Terminal Funding Pay-As-You-Go risk community

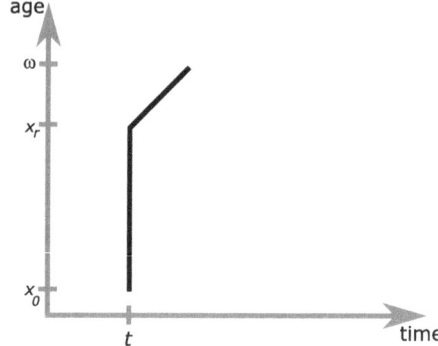

2.4 Financing the Pension Scheme

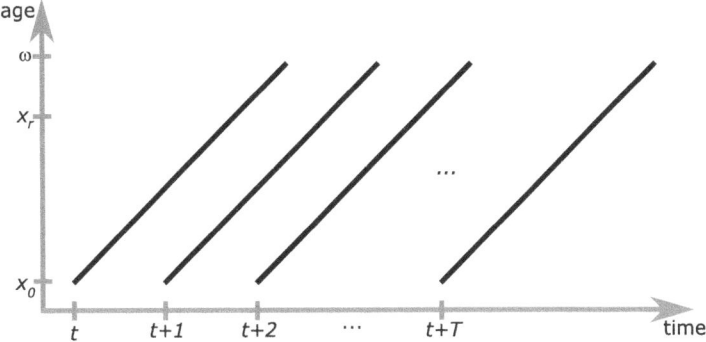

Fig. 2.6 Lexis diagram of the Collective Funded risk community

Chapter 3
Introduction to Demography

Abstract This chapter presents the basic concepts of demography and population theory that will then allow us to model the evolution of the numbers of contributors and beneficiaries of a pension scheme. The fundamental notions of stable and stationary populations are introduced. Different demographic ratios useful in pension theory are defined and illustrated, as well as the two main types of demographic risks affecting pension plans: longevity risk and population growth risk.

3.1 Lexis Diagram

As Cox (*Demography*, 1976) states, "demography is the study by statistical methods of human populants involving the measurement of their size, growth and diminution, the proportions living within some areas, being born, marrying, having children or dying and the related functions of fertility, nuptiality and mortality". We will thus be concerned here with the evolution over time of the size of a given population, generally analyzed by age. The two main analytical variables are therefore time (t) and age (x).

The Lexis diagram allows us to visualize these two variables in a two-dimensional plane, as well as the progression of a population, as shown in Fig. 3.1.

An individual in this plane is characterized by

- an entry point (x_0, t_0): depending on the considered problem, x_0 can represent birth ($x_0 = 0$); it can also represent entry into service in a company if one is interested in the evolution of its workforce,
- an exit point $(x_n, t_0 + x_n - x_0)$: depending on the case, x_n may be the age of death or the retirement age denoted by x_r. When x_n refers to the last age of a mortality table, it is denoted by ω,
- a diagonal joining these two points, called a lifeline.

Different contexts lead to different interpretations of the Lexis diagrams. For example, one could take into account immigration and emigration: entry and exit points of the Lexis diagram then represent arrival into and departure from the considered country.

Fig. 3.1 Elements of a general Lexis diagram

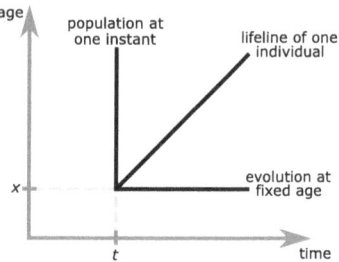

A vertical line in this diagram represents the entire population present at one instant. A horizontal line in this diagram represents the evolution over time of a fixed age group.

3.2 Population Models

A population model consists of describing the size of a population as a function of two variables: time and age.

3.2.1 A Discrete Model

We assume that the time horizon is discrete, of the form

$$T = \{t_0, t_0 + 1, t_0 + 2, \ldots t_1\}.$$

We therefore observe the population, for example, every year or every month at fixed dates.

At the age level, the scale is also discretized (by rounding the exact age to the nearest 6 months, for example):

$$X = \{x_0, x_0 + 1, x_0 + 2, \ldots, x_n\},$$

where x_0 is the first age, and x_n is the last age. We then denote by $L(x, t)$ (with $x \in X, t \in T$) the population function giving, at the date t, the size of the population aged x.

The total population at the date t is defined as

$$N(t) = \sum_{x=x_0}^{x_n} L(x, t).$$

3.2 Population Models

Fig. 3.2 Age pyramid

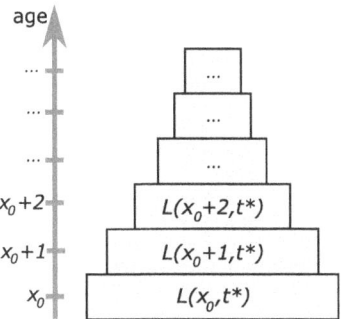

The function L of two variables allows a triple analysis:

- for $t = t^*$ fixed: L is seen as a function of age:

$$\{L(x_0, t^*), L(x_0 + 1, t^*), \ldots . L(x_n, t^*)\}.$$

This sequence is commonly represented in the form of a graph called an age pyramid (Fig. 3.2).

- for $x = x^*$ fixed: L is seen as a function of time:

$$\{L(x^*, t_0), L(x^*, t_0 + 1), \ldots . L(x^*, t_1)\},$$

giving the evolution of a fixed age group in terms of number of individuals.

- for x and t linked by $x = x^* + t - t^*$: L is seen as a function of a cohort:

$$\{L(x^*, t^*), L(x^* + 1, t^* + 1), L(x^* + 2, t^* + 2), \ldots\},$$

allowing us to track the evolution of an initial age cohort x^* at time t^* ($x^* \in X, t^* \in T$).

Population dynamics attempts to understand the evolution of population sizes based on two explanatory phenomena:

- entries into the population (birth, immigration, hiring, etc.). We denote by $E(x, t)$ (with $x \in X, t \in T$) the function giving the number of new entrants at time t at age x.
- exits from the population (death, emigration, resignation, etc.). We denote by $p_t(x, x')$ (with $x, x' \in X, t \in T$) the probability that an individual belonging to the population at age x and at time t will still be there at the age x'. These probabilities can be generated by one cause of exit (e.g. death) or by several causes of exit (e.g. death and disability). Here we will consider the total probability of survival, while in Sect. 3.6 we will study multi-decrement tables with several causes of exit.

These different population functions are linked by a relation called the renewal equation, based on a recursive formula for a given cohort:

$$L(x+h, t+h) = L(x,t) \cdot p_t(x, x+h) + \sum_{k=1}^{h} E(x+k, t+k)$$
$$\cdot p_{t+k}(x+k, x+h), \quad (3.1)$$

expressing that to be present in the population in h years, one must either have been there initially, and have survived, or one must have entered in the meantime and also survived.

An important special case is where entries are made at a single age (e.g. births at age $x = 0$). In this case, we have

$$E(x,t) = \begin{cases} E(t) & x = x_0, \\ 0 & x \neq x_0. \end{cases}$$

Equation (3.1) can then be written

$$L(x,t) = E(t - (x - x_0)) \cdot p_{t-(x-x_0)}(x_0, x).$$

If we also assume that the probabilities of survival are independent of time, we obtain

$$L(x,t) = E(t - (x - x_0)) \cdot p(x_0; x). \quad (3.2)$$

3.2.2 A Continuous Model

Even if it is natural to use discretization (as is usually done in practice), the variables "time" and "age" are of course by essence continuous. In this context, it is easy to generalize the concepts introduced previously.

We will limit ourselves here to the case of entries at a single entry age, assumed to correspond to birth ($x_0 = 0$). We then define the population density of age x at time t by a function of two variables:

$$\{\ell(x,t); x \in [0, \omega]; t \in [t_0, t_1]\}.$$

Given this function, we may calculate the evolution of the population by integration:

- $L(x_0; x_1; t) = \int_{x_0}^{x_1} \ell(x,t)dx$: population between ages x_0 and x_1, at time t;
- $\int_{t'}^{t''} \ell(x,t)dt$: population reaching age x between time t' and time t'';
- $N(t) = \int_0^{\omega} \ell(x,t)dx$ = entire population at time t.

3.2 Population Models

In this population, entries and exits are controlled by the entry and survival processes:

- entry process: $\{b(t); t \in [t_0, t_1]\}$, and $\int_{t'}^{t''} b(t)dt$ represents the number of births between time t' and time t'';
- survival process: $\{s(x, t), x \in [0, \omega]; t \in [t_0, t_1]\}$.

The relationships between these different functions are

$$\ell(0, t) = b(t), \tag{3.3}$$

$$\ell(x, t) = b(t - x)s(x, t). \tag{3.4}$$

The instantaneous exit force is also introduced, where it exists:

$$\mu(x, t) = -\frac{1}{s(x, t)} \frac{\partial}{\partial x} s(x, t). \tag{3.5}$$

Population density also satisfies, as in the discrete case, recursive equations:

1. discrete form:

$$\begin{aligned}
\ell(x + h, t + h) &= b(t + h - (x + h)) \cdot s(x + h, t + h) \\
&= b(t - x) \cdot s(x + h, t + h) \\
&= b(t - x) \cdot s(x, t) \cdot \frac{s(x + h, t + h)}{s(x, t)} \\
&= \ell(x, t) \cdot p_t(x; x + h),
\end{aligned}$$

where $p_t(x, x + h) = \frac{s(x+h,t+h)}{s(x,t)}$ represents the conditional probability of still being alive at age $x + h$ given that one was alive at time t and age x,

2. Continuous form:

$$\ell(x, t) = b(t - x) \cdot s(x, t)$$

taking the derivative, we obtain

$$\frac{\partial \ell}{\partial x} = -b'(t - x) \cdot s(x, t) + b(t - x) \cdot \frac{\partial}{\partial x} s(x, t),$$

$$\frac{\partial \ell}{\partial t} = b'(t - x) \cdot s(x, t) + b(t - x) \cdot \frac{\partial}{\partial t} s(x, t).$$

By adding these two partial derivatives, we have

$$\frac{\partial \ell}{\partial x} + \frac{\partial \ell}{\partial t} = b(t - x) \left[\frac{\partial s}{\partial x} + \frac{\partial s}{\partial t} \right].$$

In particular, if the survival process is independent of time,

$$\frac{\partial \ell}{\partial x} + \frac{\partial \ell}{\partial t} = b(t-x) \cdot s(x,t) \cdot \frac{1}{s(x,t)} \frac{\partial s}{\partial x}$$

or, considering Eqs. (3.4) and (3.5),

$$\frac{\partial \ell}{\partial x} + \frac{\partial \ell}{\partial t} = -\ell(x,t) \cdot \mu(x,t).$$

3.3 Stable Populations

Among all populations, demography is particularly interested in certain particular populations; these populations will have interesting properties in terms of evolution of pension cost.

3.3.1 Stable Discrete Populations

A population is said to be stable when its population function meets the separation criterion

$$L(x,t) = \varphi(t)\chi(x), \qquad (3.6)$$

where φ is a function only of time and χ is a function only of age.

Such populations have an age pyramid of constant shape over time; in particular the following ratios are constant over time:

1. ratio between the size of two age groups:

$$\frac{L(x_2,t)}{L(x_1,t)} = \frac{\chi(x_2)\varphi(t)}{\chi(x_1)\varphi(t)} = \frac{\chi(x_2)}{\chi(x_1)},$$

2. ratio between the size of two subgroups of the population:

$$\frac{\sum_{x=x_3}^{x_4} L(x,t)}{\sum_{x=x_1}^{x_2} L(x,t)} = \frac{\sum_{x=x_3}^{x_4} L(x,t')}{\sum_{x=x_1}^{x_2} L(x,t')} = \frac{\sum_{x=x_3}^{x_4} \chi(x)}{\sum_{x=x_1}^{x_2} \chi(x)}. \qquad (3.7)$$

3.3.2 Stationary Discrete Populations

A population is said to be stationary when its population function is independent of time:

$$L(x, t) = \chi(x).$$

In this case, not only does the form of the population remain unchanged, but its size also remains constant.

3.3.3 How to Obtain Stable Populations

The following assumptions can be used in order to try to find stationary populations:

1. the entries into the population happen at a uniform age, denoted x_0,
2. the probabilities of survival are independent of time.

In this case, the population function can be written (see Eq. (3.2)):

$$L(x, t) = E(t - (x - x_0)) \cdot p(x_0; x).$$

In general, this expression does not meet the separation criterion of Eq. (3.6). For example, if the entry function is a linear function, we have

$$E(t - (x - x_0)) = K \cdot (t - (x - x_0)),$$
$$L(x, t) = K(t - (x - x_0)) \cdot p(x_0; x),$$

which is not separable.

On the other hand, if the entry function is an exponential function, we obtain

$$E(t - (x - x_0)) = A \cdot e^{\rho(t-(x-x_0))},$$
$$L(x, t) = A \cdot e^{\rho(t-(x-x_0))} \cdot p(x_0; x) = \left(A \cdot e^{\rho t}\right) \cdot \left(e^{-\rho(x-x_0)} \cdot p(x_0; x)\right), \quad (3.8)$$

which is separable.

The total population is given in this case by

$$N(t) = \sum_{x=x_0}^{x_n} L(x, t) = e^{\rho t} \cdot \left(\sum_{x=x_0}^{x_n} A \cdot e^{-\rho(x-x_0)} \cdot p(x_0; x)\right) = C \cdot e^{\rho t},$$

i.e. by an exponential total population.

In particular,

- if $\rho > 0$, we have an increasing stable population,
- if $\rho = 0$, we have a stationary population,
- if $\rho < 0$, we have a decreasing stable population.

3.3.4 Stable Continuous Populations

These notions of stability can easily be generalized in a continuous framework.

An example of such a continuous stable population with a uniform entry age $x_0 = 0$ (cf. Eq. (3.4)) is given by

$$\ell(x, t) = b(t - x) \cdot s(x, t) = b \cdot e^{R(t-x)} \cdot s(x),$$

where R is a constant, i.e. the case where the survival process is independent of time, and the entry process is exponential.

In this model, the total population is given by

$$N(t) = \int_0^\omega \ell(x, t) dx = \int_0^\omega b e^{R(t-x)} s(x) dx = e^{Rt} \int_0^\omega b e^{-Rx} s(x) dx = e^{Rt} N_0.$$

For $R = 0$, we have a stationary population.

3.3.5 An Example of a Continuous Unstable Population

For a stable population with uniform entry age, the function N solves the following linear differential equation:

$$\frac{dN}{dt}(t) = RN(t).$$

Non-stationary models can be developed, such as the *logistic model*, where it is assumed that the growth rate R decreases linearly as the population increases, taking into account, for example, environmental limits:

$$\frac{dN}{dt}(t) = R(N)N(t),$$

where $R(N) = r - bN$.

3.3 Stable Populations

The differential equation then becomes

$$\frac{dN}{dt}(t) = rN(t) - bN^2(t).$$

In order to solve this differential equation, let us define

$$\varphi(t) = \frac{1}{N(t)}.$$

We then obtain

$$\varphi'(t) = -\frac{N'(t)}{N^2(t)} = -\frac{N(t)r - bN^2(t)}{N^2(t)} = -\frac{r}{N(t)} + b = -r\varphi(t) + b.$$

It is easy to check that the following function is a solution of this differential equation:

$$\varphi(t) = \frac{b}{r}\left[1 + Ke^{-rt}\right],$$

where K is a constant. Indeed, by differentiating the function φ, we get

$$\varphi'(t) = -bKe^{-rt}.$$

On the other hand, considering the form of φ, one has

$$\frac{r}{b}\varphi(t) - 1 = Ke^{-rt}.$$

By substituting, we obtain

$$\varphi'(t) = -b\left(\frac{r}{b}\varphi(t) - 1\right) = -r\varphi(t) + b.$$

Coming back to the original function N:

$$N(t) = \frac{r}{b}\frac{1}{1 + Ke^{-rt}}.$$

The value of the constant K can be obtained from the initial condition

$$N(0) = \frac{r}{b}\frac{1}{1 + K},$$

$$K = \frac{r}{bN(0)} - 1.$$

Finally, by replacing, we have

$$N(t) = \frac{r}{b} \frac{1}{1 + \left(\frac{r}{bN(0)} - 1\right)e^{-rt}},$$

which can still be written, by defining $N^* = \frac{r}{b}$,

$$N(t) = N(0) \frac{N^*}{N(0)\left(1 - e^{-rt}\right) + N^* e^{-rt}}.$$

The population gradually changes from an initial size of $N(0)$ to an asymptotic size of N^*. In particular,

$$\lim_{t \to \infty} N(t) = N^*.$$

Unlike increasing stable populations, the number of people is bounded here.

3.4 Demographic Ratios

Let us now derive some demographic ratios, which are very useful quantities in pension theory. These ratios will be defined in a discrete time setting, but they can be straightforwardly generalized to a continuous time setting.

It is traditional in social security theory to divide the overall population into three age subgroups:

- the young people: $x \in \{0, 1, \ldots, x_a - 1\}$,
- the active people: $x \in \{x_a, x_a + 1, \ldots, x_r - 1\}$,
- the retired people: $x \in \{x_r, x_r + 1, \ldots, \omega\}$.

Note that we focus on *demographic* ratios, as we define these groups only on age criteria. Some authors focus on *economic* ratios, considering that active people should only include working individuals, i.e. excluding unemployed people from this category.

The following different ratios can then be introduced, assuming a discrete population model:

1. The dependence ratio is the ratio between the number of retired people and the number of active people:

$$D(t) = \frac{\sum_{x=x_r}^{\omega} L(x, t)}{\sum_{x=x_a}^{x_r - 1} L(x, t)}. \tag{3.9}$$

It measures the dependence of the retired population on the active population; it plays an important role in pension theory.

2. The potential support ratio is the inverse of the dependence ratio and measures the number of active people per retiree:

$$F(t) = \frac{1}{D(t)}.$$

3. The aging intensity coefficient measures the proportion of the retired population over a certain age (e.g. 80 years). Defining $x_r < x_v < \omega$,

$$G(t) = \frac{\sum_{x=x_v}^{\omega} L(x,t)}{\sum_{x=x_r}^{\omega} L(x,t)}.$$

4. The youth ratio is the ratio between the number of young people and the number of active people:

$$H(t) = \frac{\sum_{x=0}^{x_a-1} L(x,t)}{\sum_{x=x_a}^{x_r-1} L(x,t)}.$$

5. The total dependency ratio is the sum of the dependence ratio and the youth ratio. It measures the overall degree of dependence of the inactive population ("youth" and "retirees") on the active population:

$$I(t) = D(t) + H(t) = \frac{\sum_{x=0}^{x_a-1} L(x,t) + \sum_{x=x_r}^{\omega} L(x,t)}{\sum_{x=x_a}^{x_r-1} L(x,t)}.$$

In the case of a stable population, all these demographic ratios are constant over time since in this type of population, the ratios between age groups do not change over time (see Eq. (3.7)). This explains the theoretical interest of this type of population in social security issues.

3.5 Demographic Risks

The general renewal relationship of a population shows that two processes play a fundamental role in the evolution over time of the size of different age groups:

- the renewal of the population modeled by its entry function,
- the lifetime of the population modeled by survival probabilities.

These two different processes will generate two demographic risks on a given population, both influencing the financial evolution of a pension plan: the *population growth risk* and the *longevity risk*. The first corresponds, for example, to a decrease in the birth rate, while the second is induced by an increase in longevity. It is therefore easy to understand the effects on a Pay-As-You-Go pension scheme: the first phenomenon is likely to reduce the amount of received contributions, as the working population will gradually decrease, while the second phenomenon will lead to an increase in costs, as benefits will have to be paid for a longer period.

In the rest of this section, we will assume that the population has a single entry age. The population function is then given explicitly by Eq. (3.2):

$$L(x,t) = E\left(t - (x - x_0)\right) \cdot p_{t-(x-x_0)}(x_0, x).$$

3.5.1 Longevity Risk

A classic actuarial tool for measuring and highlighting longevity risk is the *life expectancy*; it can be defined as the mathematical expectation at a given age of the number of years remaining to be lived. If defined at retirement age, it represents for a given cohort the average number of years of pension benefit payments.

In the discrete model considered here it is a function of two variables given by

$$e_x(t) = \text{life expectancy for an individual of age } x \text{ at time } t = \sum_{y=x+1}^{\omega} p_t(x, y).$$

This function is sometimes called *prospective life expectancy* since it uses future survival probabilities. It implicitly takes into account the expected evolution of mortality over time.

If we also assume that the probabilities of survival are stationary over time, we obtain the traditional actuarial life expectancy, which is a function of age only and is often calculated using the probabilities observed at a given time (see the technical appendix for details):

$$e_x = \sum_{y=x+1}^{\omega} p(x,y) = \sum_{y=x+1}^{\omega} {}_{y-x}p_x.$$

Two life expectancies are generally used in practice:
- life expectancy at birth ($x = 0$),
- life expectancy at retirement ($x = x_r$).

3.5 Demographic Risks

Life expectancy can be naturally associated with the annuity price (see the technical appendix for details). Let us consider for this the price of a revaluated life annuity (in arrears):

$$a_x^{(i,g)} = \sum_{y=x+1}^{\omega} \frac{(1+g)^{y-x}}{(1+i)^{y-x}} {}_{y-x}p_x,$$

where i is the discount rate, g is the revaluation rate and the ${}_\bullet p_x$ are the survival probabilities.

Considering the particular case of a revaluation rate equal to the discount rate, we obtain

$$a_x^{(i,i)} = \sum_{y=x+1}^{\omega} {}_{y-x}p_x = e_x.$$

3.5.2 Demographic Growth Risk

Population growth risk is related to the rate of active people renewal and is therefore directly generated by the population entry function, denoted $E(t)$. Entries into the population generally come from either births or migrations. Various coefficients can be used to measure the importance of these inputs. One example of this is the fertility rate for births, defined by the average number of children per woman of childbearing age. It is generally considered in demography that this index must be at least 2.1 to ensure the simple renewal of the population.

3.5.3 A Two-Period Model

We present here a simple two-period model showing the effects of the two basic demographic risks (longevity and growth) on a pension plan. So we will obtain a quantitative comparison between the main financing methods seen in Sect. 2.4.2: Pay-As-You-Go and Full Funding.

In this simple stylized model, each individual lives two periods in his or her life: one as an active person and the other as a retiree. Individuals enter the population at age $x_r - 1$, retire at age x_r and die at age $x_r + 1$.

An average salary is received at age $x_r - 1$, from which a pension contribution is computed in order to finance the pension paid at age x_r. It is assumed that the pension scheme is a Defined Benefit scheme based on final salary: the pension is equal to a proportion δ of the salary of the working population. The demographic variables satisfy the following recursive relations:

- active population at time t:

$$L(x_r - 1, t) = E(t) = E(t-1)(1 + d_{t-1}),$$

- retired population at time t:

$$L(x_r, t) = L(x_r - 1, t-1) p_{t-1} = E(t-1) p_{t-1},$$

where

- $E(t)$ = entry in the population at time t,
- d_t = active population growth rate at time t,
- p_t = survival probability of active people at time t.

On the other hand, the following financial quantities are introduced:

- g_t = salary growth rate at time t,
- i_t = interest rate at time t.

All working people receive the same salary given at time t by

$$\overline{S}(t) = S_0 \prod_{j=1}^{t} (1 + g_j).$$

Retired people receive a pension at time t which is assumed to correspond to a fixed percentage of the salary of working people:

$$\overline{R}(t) = \delta \overline{S}(t).$$

3.5.3.1 Pay-As-You-Go

In accordance with the definitions introduced in Sect. 2.4, the Pay-As-You-Go contribution rate is simply expressed as the ratio between the pension costs payable at time t and the sum of the salaries on which the rate applies at this instant:

$$\begin{aligned}
\pi_{PA}(t) &= \text{contribution rate in PAYG at time } t \\
&= \frac{\text{pension costs at time } t}{\text{salaries of the active people at time } t} \\
&= \frac{\delta \overline{S}(t) L(x_r, t)}{\overline{S}(t) L(x_r - 1, t)} \\
&= \delta p_{t-1} \frac{1}{(1 + d_{t-1})}.
\end{aligned} \quad (3.10)$$

3.5 Demographic Risks

This formula for the PAYG contribution rate illustrates the the exposure of the pension scheme to the two demographic risks:

- the longevity risk through the probability of survival p_{t-1}: the contribution rate increases linearly with the probability of survival of retirees,
- the demographic growth risk through the growth coefficient d_{t-1}: the contribution rate is inversely proportional to the growth rate of the active population.

3.5.3.2 Full Funding

In this case, the contribution paid by the active people at time t will allow them to fund their own retirement at time $t+1$. We therefore have a balanced relationship which expresses this time in terms of discounted values and which (also in PAYG) takes into account an adjustment of pensions at the same rate as salaries, denoting by $\pi_{FF}(t)$ the contribution rate in full funding:

$$\pi_{FF}(t)\overline{S}(t)L(x_r - 1, t) = \delta(1 + g_t)\overline{S}(t)L(x_r, t+1)\frac{1}{(1+i_t)}$$

$$\Rightarrow \quad \pi_{FF}(t) = \delta p_t \frac{(1+g_t)}{(1+i_t)}. \tag{3.11}$$

Now the contribution rate no longer depends on population growth; on the other hand, longevity risk is still present; moreover, financial market risks appear. Formulas (3.10) and (3.11) also allow a first comparison between PAYG and fully funded systems in the context of the simplified economy considered here. The contribution rates for Pay-As-You-Go and fully funded systems will be the same if the following equation involving demographic and financial parameters holds:

$$\frac{p_t}{p_{t-1}}(1 + d_{t-1}) = \frac{1 + i_t}{1 + g_t}.$$

In particular, if all the parameters are constant over time, we obtain

$$1 + d = \frac{1+i}{1+g}, \tag{3.12}$$

expressing that the population growth rate corresponds to the interest rate adjusted by the salary growth rate. This fundamental relationship will be generalized in Chap. 4 in a multi-period model.

3.6 Multiple-Decrement Tables

The demographic functions studied above are influenced in particular by a survival function denoted $p_t(x, x')$ (the probability of survival for an individual at age x' who is present at time t at age x). This survival process can be considered globally as generated by one exit cause; it is sometimes necessary to consider and explain several exit causes acting simultaneously on the population.

In traditional life insurance, for example, only one cause of exit is considered: death. In disability insurance, we are interested in two causes of exit: death and disability. In the management of a pension plan, several causes of exit can be taken into account: death, disability, turnover; if the retirement age is variable, retirement can also be considered as a cause for leaving the working population.

In this context, multiple-decrement tables make it possible to link each of the exit probabilities of the various factors to the total survival probabilities of the population. This requires the introduction of dependent and independent probability concepts.

For example, let us first consider a model with two causes of exit (e.g. death and disability); we assume that the population is stable. We therefore define at the general level of the population

$$p(x, x+1) =_1 p_x = \frac{\ell_{x+1}}{\ell_x} = \text{total probability of survival between ages } x \text{ and } x+1.$$

If two causes of elimination affect the active population, we define

$$d_x = \ell_x - \ell_{x+1} = d_x^{(1)} + d_x^{(2)},$$

where $d_x^{(1)}$ is the number of exits induced by the first elimination cause and $d_x^{(2)}$ is the number of exits induced by the second elimination cause.

The total probability of annual exit can also be split as

$$q_x = \frac{d_x}{\ell_x} = \frac{d_x^{(1)}}{\ell_x} + \frac{d_x^{(2)}}{\ell_x} = q_x^{*(1)} + q_x^{*(2)},$$

where $q_x^{*(1)}$ and $q_x^{*(2)}$ are called annual dependent exit probabilities. They are said to be *dependent* because they are calculated on a population which is simultaneously exposed to both causes.

The total probability of survival becomes

$$p_x = 1 - q_x = 1 - q_x^{*(1)} - q_x^{*(2)}.$$

These dependent probabilities can be linked to independent probabilities of exit of each of the causes (probabilities of each of the causes considered separately), denoted $q_x^{(1)}$ and $q_x^{(2)}$, such that

$$p_x = 1 - q_x = \left(1 - q_x^{(1)}\right)\left(1 - q_x^{(2)}\right).$$

3.6 Multiple-Decrement Tables

In order to calculate these independent probabilities, the number of people exposed to one cause should be adjusted according to the exits caused by the other cause. Assuming a uniform distribution throughout the year, we then have

$$q_x^{(1)} \cong \frac{d_x^{(1)}}{\ell_x - \frac{1}{2}d_x^{(2)}} = \frac{q_x^{*(1)}}{1 - \frac{1}{2}q_x^{*(2)}},$$

$$q_x^{(2)} \cong \frac{d_x^{(2)}}{\ell_x - \frac{1}{2}d_x^{(1)}} = \frac{q_x^{*(2)}}{1 - \frac{1}{2}q_x^{*(1)}}.$$

By inverting these relationships we can obtain the dependent probabilities as functions of the independent probabilities:

$$q_x^{*(1)} \cong q_x^{(1)}\left(1 - \frac{1}{2}q_x^{(2)}\right), \qquad (3.13)$$

$$q_x^{*(2)} \cong q_x^{(2)}\left(1 - \frac{1}{2}q_x^{(1)}\right). \qquad (3.14)$$

The dependent probabilities are therefore lower than the corresponding independent probabilities.

This ranking between dependent and independent probabilities can be intuitively explained. Dependent probabilities are expressed as the ratio between the number of exits for a given cause and the total size of the population at the beginning of the year. However, because of the presence of other causes of exit during the year, this measure overestimates the size of the population exposed to that cause throughout the year (and therefore, in a sense, the probabilities it gives are "too small"): a part of the population will not be exposed to a given cause of exit because they already exit for another cause! Independent probabilities correct that bias and are therefore higher.

The same methodology can be used in the presence of more than two causes of exit. For n causes of exits, the following approximation is commonly used, which generalizes Formula (3.13):

$$q_x^{*(1)} \cong q_x^{(1)} \prod_{j=2}^{n}\left(1 - q_x^{(j)}/2\right).$$

This gives, for instance, for three causes (e.g. death, disability and resignation),

$$q_x^{*(1)} \cong q_x^{(1)}\left(1 - \frac{1}{2}\left(q_x^{(2)} + q_x^{(3)}\right) + \frac{1}{4}q_x^{(2)}q_x^{(3)}\right).$$

Chapter 4
General Funding Systems

Abstract This chapter deals more formally with the different funding methods for pension schemes. After explaining the general equation of actuarial equilibrium, we review the main families of methods. Finally, a simple macroeconomic model makes it possible to compare two extreme funding methods and to state the famous social security paradox.

4.1 General Actuarial Equilibrium Equation of a Pension Scheme

4.1.1 Actuarial Equilibrium and Funding Methods

The equilibrium equation of a contributive pension scheme is based on the principle of actuarial equivalence between benefits and contributions: future benefits must be funded either by future contributions or by any existing initial reserve. In pension theory, this equivalence is not generally expressed at the level of each individual (we do not necessarily seek equilibrium at the individual level) but at the level of some (sub)-population. So we have the following relationship:

$$\sum_{\text{individuals}} \text{discounted value of contributions} + \text{initial reserve}$$

$$= \sum_{\text{individuals}} \text{discounted value of benefits.} \quad (4.1)$$

In order to introduce the basic equations linking reserves, benefits and contributions, we assume that the system starts at time $t = 0$ and has an initial reserve $V(0)$. We use the following notation:

- $\pi(s)$ = the average contribution rate of the scheme at time s,
- $\overline{S}(s)$ = the average salary of the working population at time s,
- $SS(s) = \sum_{x=x_0}^{x_r - 1} L(x, s)\overline{S}(s)$ = the total payroll at time s,

© The Author(s), under exclusive license to Springer Nature Switzerland AG 2025
P. Devolder, S. de Valeriola, *Actuarial Pension Funding Theory*, Springer Actuarial Textbooks, https://doi.org/10.1007/978-3-031-85268-8_4

- $\overline{R}(s)$ = the average pension benefit paid at time s,
- $SR(s) = \sum_{x=x_r}^{\infty} L(x,s)\overline{R}(s)$ = the total benefits paid at time s,
- i = the rate of return on the reserves over one period.

We will assume here that contributions and benefits are paid annually at the start of each year.

We can then obtain a first expression of the scheme's reserves, *in retrospective form* (we will assume here that salaries and benefits are payable in advance):

$$V(t) = V(0) \cdot (1+i)^t + \sum_{s=0}^{t-1} (\pi(s)SS(s) - SR(s)) \cdot (1+i)^{t-s}. \tag{4.2}$$

This equation makes it possible to obtain a second one, called the recurrence formula for reserves:

$$V(t+1) = (V(t) + \pi(t)SS(t) - SR(t))(1+i) \tag{4.3}$$

or, in differential form,

$$\Delta V(t+1) = V(t+1) - V(t) = i \cdot V(t) + (\pi(t)SS(t) - SR(t))(1+i),$$

expressing that the increase in the scheme's reserves comes from interest on existing reserves, and from the excess of contributions over benefits.

The equation of actuarial equilibrium (4.1) finally allows us to express the reserves in *prospective form*:

$$\text{Reserve} = \text{discounted value of future benefits}$$
$$\quad\quad\quad - \text{ discounted value of future contributions,}$$

$$V(t) = \sum_{s=t}^{\infty} SR(s)(1+i)^{-(s-t)} - \sum_{s=t}^{\infty} \pi(s)SS(s)(1+i)^{-(s-t)}. \tag{4.4}$$

The actuarial balance between benefits and contributions is then expressed simply by writing that the retrospective and prospective forms (4.2) and (4.4) of the scheme's reserve must be equal at any time:

$$V(0)(1+i)^t + \sum_{s=0}^{t-1} (\pi(s)SS(s) - SR(s))(1+i)^{t-s}$$
$$= \sum_{s=t}^{\infty} SR(s)(1+i)^{-(s-t)} - \sum_{s=t}^{\infty} \pi(s)SS(s)(1+i)^{-(s-t)}. \tag{4.5}$$

4.1 General Actuarial Equilibrium Equation of a Pension Scheme

This formula expresses the equality between the retrospective reserve accumulated at time t and the prospective reserve at time t. In particular at time 0, we can write:

$$V(0) + \sum_{s=0}^{\infty} \pi(s) SS(s)(1+i)^{-s} = \sum_{s=0}^{\infty} SR(s)(1+i)^{-s},$$

meaning that the initial reserves plus the present value of future contributions must be equal to the present values of future benefits.

In a Defined Benefit scheme, the vector $\{SR\}$ is fixed; a funding method is then a vector $\{\pi\}$ satisfying this equation. We remark that there are an infinite number of solutions for this vector. Each solution corresponds to a particular funding method. In contrast, in a Defined Contribution scheme, the vector $\{\pi\}$ is fixed and the benefits granted by the scheme must be such that Eq. (4.5) is satisfied.

Different basic funding methods can be identified.

4.1.1.1 Limit Case n°1

We impose the constraint of having zero reserves at all times (we assume in this case that we start with a zero reserve level): $V(t) = 0 \ \forall t$. The recurrence Eq. (4.3) shows that we must then have

$$\pi(t) \cdot SS(t) - SR(t) = 0 \quad \forall t.$$

In a Defined Benefit scheme, this leads to the following contribution rate:

$$\pi(t) = \frac{SR(t)}{SS(t)}. \tag{4.6}$$

In a Defined Contribution scheme, the same equation leads to the following benefits:

$$SR(t) = \pi(t) \cdot SS(t) \quad \text{or} \quad \overline{R}(t) = \frac{\pi(t) \cdot SS(t)}{\sum_{x=x_r}^{\omega} L(x,t)}.$$

This is the Pay-As-You-Go method.

4.1.1.2 Limit Case n°2

We impose the constraint of no longer having to pay any future contribution: $\pi(t) = 0 \ \forall t$. The initial reserve must therefore be sufficient to fund all future benefits of the scheme.

$$V(0) = \sum_{s=0}^{\infty} SR(s)(1+i)^{-s} = \sum_{s=0}^{\infty} \sum_{x=x_r}^{\omega} L(x,s) \overline{R}(s)(1+i)^{-s}.$$

This is the "initial funding" method. It is obviously the method that generates the largest reserves.

4.1.1.3 Case n°3

We can introduce another method by imposing the constraint of having a contribution rate that is constant over time, with a zero initial reserve: $\pi(t) = \pi \ \forall t$,

$$\pi = \frac{\sum_{s=0}^{\infty} SR(s)(1+i)^{-s}}{\sum_{s=0}^{\infty} SS(s)(1+i)^{-s}} = \frac{\sum_{s=0}^{\infty} \sum_{x=x_r}^{\omega} L(x,s)\overline{R}(s)(1+i)^{-s}}{\sum_{s=0}^{\infty} \sum_{x=x_0}^{x_r-1} L(x,s)\overline{S}(s)(1+i)^{-s}}.$$

This is the average general premium method.

Generally speaking, in a Defined Benefit scheme, an *acceptable* funding method is defined as a vector $\{\pi\}$ satisfying the equilibrium Eq. (4.5) such that the level of the reserves is always between Pay-As-You-Go and Initial Funding:

$$0 \leq V(t) \leq \sum_{s=t}^{\infty} \sum_{x=x_r}^{\omega} L(x,s)\overline{R}(s)(1+i)^{-(s-t)}.$$

We therefore reject methods which, although actuarially balanced, can generate negative reserves at some point. As a counter-example, the following case can be considered, where nothing is paid until time T:

$$V(0) = 0 \ ; \quad \pi(s) = 0 \ \forall s = 0, \ldots, T-1 \quad \text{and}$$

$$\pi(T)SS(T) = \sum_{s=0}^{\infty} SR(s)(1+i)^{T-s}.$$

This last equation, reflecting a full payment as in the case of Initial Funding but made too late, illustrates the actuarial equilibrium nature of the method. Nevertheless, if benefits are to be paid before the instant T, i.e. if $SR(s) \neq 0$ for $s < T$, the reserves will be negative, and the method is therefore not acceptable.

4.1.2 Workers' Reserve and Retirees' Reserve

The total reserves $V(t)$ of a pension scheme can generally be divided into two parts:

$$V(t) = V_r(t) + V_a(t), \tag{4.7}$$

4.1 General Actuarial Equilibrium Equation of a Pension Scheme

where

$V_r(t) =$ reserves necessary to fully fund the pensions of individuals already retired at time t

$ =$ retirees' reserve

$$= \sum_{x=x_r}^{\infty} \sum_{s=t}^{t+\omega-x} L(x,s)\overline{R}(s)(1+i)^{-(s-t)},$$

$V_a(t) =$ workers' reserve

$ = V(t) - V_r(t).$

The retirees' reserve does not depend on the choice of the funding method; it can be seen as a general theoretical level. By definition, the workers' reserve is the excess of total reserves over the retirees' reserve. Its level therefore depends on the choice of funding method.

The different funding methods can then be classified according to the sign of these different reserves.

4.1.2.1 Pay-As-You-Go Method

$$V(t) = 0 \quad \text{and} \quad V_a(t) = -V_r(t) < 0$$

The scheme has no reserves. The retirees' reserve is a liability borne by the working population. The workers therefore have a debt towards the system.

4.1.2.2 Terminal Funding Method

$$V(t) = V_r(t) \quad \text{and} \quad V_a(t) = 0$$

The scheme has at all times exactly the amount needed for the retirees' reserve. The workers have no reserves, but no debt.

4.1.2.3 Fully Funded Methods

$$V(t) > V_r(t) \quad \text{and} \quad V_a(t) > 0$$

The retirees' reserve is funded and the workers have a strictly positive reserve.

4.1.3 Demographic Projections

One of the essential elements appearing in equilibrium equations is the choice of the function $L(x, t)$ modeling the population to which the actuarial equivalence relates.

Starting from the existing population at observation date t, it is common in pension models to encounter three situations.

4.1.3.1 Closed Population without Replacements

In this case,

$$E(x, s) = 0 \quad \forall s \geq t, \forall x.$$

The initial population is progressively decreasing when individuals leave. The series appearing for example in Formula (4.4) then become finite sums, since beyond the date $t + \omega$ at the latest, the population function becomes equal to zero.

4.1.3.2 Closed Population with Replacements

In this case, the entry function estimated at time t is not equal to zero, but becomes so beyond a certain limit time T:

$$\exists T > 0 \text{ such that } E(x, r) = 0 \quad \forall r \geq T, \forall x.$$

New future entries into the population are therefore introduced (over one or more generations), but the population is nevertheless closed at some point. As in the previous case, at the latest at time $T + \omega$, the population function becomes equal to zero.

4.1.3.3 Open Population

In this case, at any time $s > t$, new individuals enter the population:

$$\forall s \geq t : \begin{Bmatrix} \exists r \geq s \\ \exists x \end{Bmatrix} \text{ such that } E(x, r) \neq 0.$$

As a result, the population is projected onto an infinite horizon. This type of projection is mainly used in social security systems, with the implicit assumption of permanence. In contrast, in occupational schemes (i.e. in the second pillar), a closed population is always used, most often without replacements.

4.2 Classification of Funding Methods

We now turn to the study of the main funding methods. We will successively consider Pay-As-You-Go, Scaled Premium, Terminal Funding and Fully Funded schemes. Note that the definitions given in this section concern a Defined Benefit scheme.

4.2.1 Pay-As-You-Go Method

The Pay-As-You-Go method is based on the principle of equivalence at any time between the contributions paid by all workers and the pension benefits paid to all retirees.

We use the following notation:

$\overline{R}(t) = $ the average pension paid at time t,

$\overline{S}(t) = $ the average salary at time t,

$\pi_{PA}(t) = $ the Pay-As-You-Go contribution rate,

$x_r = $ the retirement age,

$x_0 = $ the entry age.

The contributions collected at t are equal to

$$\pi_{PA}(t)\, \overline{S}(t) \sum_{x=x_0}^{x_r - 1} L(x, t),$$

while the pension expenses at t are equal to

$$\overline{R}(t) \sum_{x=x_r}^{\omega} L(x, t).$$

Applying the principle of equivalence therefore gives

$$\pi_{PA}(t)\, \overline{S}(t) \sum_{x=x_0}^{x_r - 1} L(x, t) = \overline{R}(t) \sum_{x=x_r}^{\omega} L(x, t)$$

Table 4.1 Contribution rate of a Pay-As-You-Go scheme

		D					
		40%	44%	53%	63%	67%	69%
δ	50%	20.0%	22.0%	26.5%	31.5%	33.5%	34.5%
	70%	29.0%	30.8%	37.1%	44.1%	46.9%	48.3%

or

$$\pi_{\mathrm{PA}}(t) = \frac{\overline{R}(t)}{\overline{S}(t)} \frac{\sum_{x=x_r}^{\omega} L(x,t)}{\sum_{x=x_0}^{x_r-1} L(x,t)}$$
$$= \delta(t)\, D(t),$$

where $\delta(t)$ is the benefit ratio, measuring the ratio of retiree income to working income (relative standard of living of retirees to working population, socioeconomic parameter) and $D(t)$ is the dependence ratio (see Eq. (3.9), demographic parameter).

The Pay-As-You-Go contribution rate is therefore the product of the dependence ratio and the benefit ratio. When the benefit ratio is constant, the scheme is considered mature. Recall that in a stable population, the dependence ratio is constant. In a stable population and for a mature pension scheme, the Pay-As-You-Go contribution rate is therefore constant. In other situations, the rate will change over time.

Table 4.1 gives some examples of contribution rates for different values of the dependence ratio and the benefit ratio.

To conclude, let us summarize the main characteristics of the Pay-As-You-Go method:

1. *no reserves*: contributions of the workers are directly recycled into payment of benefits for the retirees,
2. *full inter-generational transfer*: the benefits of one generation are totally dependent on another generation,
3. *need for permanence*: the method, which does not guarantee any future guarantee for either workers or retirees, can only be conceived fairly in a philosophy of permanence of the pension system,
4. *insensitivity to inflation*: the method, by its transversal nature, allows a link between pension benefits and the evolution of salaries,
5. *insensitivity to financial markets*: the method does not lead to any form of investment, and thus does not depend on the evolution of financial markets,
6. *maximum sensitivity to demographic evolution*: the method depends strongly on the demographic evolution of the population, and especially on the balance between the numbers of workers and retirees,
7. *cost stability*: when the population is stable, and the scheme is mature, the method allows stable funding over time for the pension scheme.

4.2.2 Scaled Premium PAYG Method

The Scaled Premium method is based on the same principles as the single-period Pay-As-You-Go, but rather than asking that revenues and costs must be equal over one period, we ask that their discounted values are equal over several periods:

Discounted value of workers contributions received between t and $t + T$

$=$

Discounted value of benefits paid to retirees between t and $t + T$

Analytically, on the one hand, for the discounted value of contributions on $[t, t+T]$ we get

$$\pi_{SP}(t) \cdot \sum_{s=0}^{T} \overline{S}(t+s) \left(\sum_{x=x_0}^{x_r-1} L(x, t+s) \right) v^s,$$

where $v = 1/(1+i)$ denotes the discount factor. On the other hand, for the discounted value of the expenses on $[t, t + T]$ we obtain

$$\sum_{s=0}^{T} \overline{R}(t+s) \left(\sum_{x=x_r}^{\omega} L(x, t+s) \right) \cdot v^s.$$

By applying the principle of equivalence, we get

$$\pi_{SP}(t) = \frac{\sum_{s=0}^{T} \overline{R}(t+s) \sum_{x=x_r}^{\omega} L(x, t+s) v^s}{\sum_{s=0}^{T} \overline{S}(t+s) \sum_{x=x_0}^{x_r-1} L(x, t+s) v^s}.$$

In particular, if the population is stable, taking into account Eq. (3.6), we obtain

$$\pi_{SP}(t) = \frac{\sum_{s=0}^{T} \overline{R}(t+s) \sum_{x=x_r}^{\omega} \varphi(t+s) \chi(x) v^s}{\sum_{s=0}^{T} \overline{S}(t+s) \sum_{x=x_0}^{x_r-1} \varphi(t+s) \chi(x) v^s}$$

$$= \frac{\left(\sum_{x=x_r}^{\omega} \chi(x) \right) \left(\sum_{s=0}^{T} \overline{R}(t+s) \varphi(t+s) v^s \right)}{\left(\sum_{x=x_0}^{x_r-1} \chi(x) \right) \left(\sum_{s=0}^{T} \overline{S}(t+s) \varphi(t+s) v^s \right)}$$

$$= D \, \overline{\delta}(t),$$

where D is the dependence ratio, and $\overline{\delta}(t)$ is the weighted average benefit ratio between t and $t + T$.

If we further assume that salaries and pensions move in the same way (with a constant benefit ratio), i.e. that $\overline{R}(t+s) = \delta \overline{S}(t+s)$, then we have

$$\pi_{SP}(t) = D\,\delta$$

and we find the single-period Pay-As-You-Go method again.

The use of the Scaled Premium Pay-As-You-Go method, which consists in leveling the rates resulting from the single-period Pay-As-You-Go method, therefore makes sense only when the population is not stable or when the benefit ratio is not constant.

It should be noted that the Scaled Premium Pay-As-You-Go method may not always be used because it may result in negative reserves. Indeed, if at the start of the scheme, and in the absence of pre-existing reserves, we have

$$\pi_{SP}(0) < \pi_{PA}(0) = \frac{SR(0)}{SS(0)},$$

the retrospective reserve formula (4.2) gives

$$V(1) = (\pi_{SP}(0)SS(0) - SR(0)) \cdot (1+i) < \left(\frac{SR(0)}{SS(0)}SS(0) - SR(0)\right) \cdot (1+i) = 0.$$

The Scaled Premium method is useful when there is significant costs growth in the single-period Pay-As-You-Go model. The Scaled Premium method then allows the rate increase to be pre-funded and leveled over a fixed period of time.

4.2.3 Variant: Scaled Premium Method with Buffer Fund

The Scaled Premium Method generates some provisions during the period of leveling $(t, t+T)$ but normally the provisions are exhausted at the end of this period (at time $t+T+1$). We could prefer to maintain a minimum level of reserve, also at the end of the leveling period, corresponding for instance to a multiple of the mean annual amount of benefits to pay (safety buffer fund). The actuarial equivalence over the period $(t, t+T)$ then becomes:

Reserve at time t

+Discounted value of workers contributions received between t and $t+T$

$$=$$

Discounted value of benefits paid to retirees between T and $t+T$

+Discounted target reserve at time $t+T+1$.

4.2 Classification of Funding Methods

The contribution rate is then

$$\pi_{\text{SPB}}(t) = \frac{\sum_{s=0}^{T} \overline{R}(t+s) \sum_{x=x_r}^{\omega} L(x,t+s)v^s + \left(V(t+T+1)v^{T+1} - V(t)\right)}{\sum_{s=0}^{T} \overline{S}(t+s) \sum_{x=x_0}^{x_r-1} L(x,t+s)v^s},$$

where

$V(t) = $ current reserve at time t,

$V(t+T+1) = $ target reserve at time $t+T+1$.

Let us remark that for $T = 0$, we obtain a variant of the pure Pay-As-You-Go method, where we introduce in the annual Pay-As-You-Go some form of reserving.

The corresponding contribution rate is then given by

$$\pi_{\text{PAB}}(t) = \frac{\overline{R}(t) \sum_{x=x_r}^{\omega} L(x,t) + (V(t+1)v - V(t))}{\overline{S}(t) \sum_{x=x_0}^{x_r-1} L(x,t)}.$$

Finally let us mention that, like the basic Scaled Premium Method, this variant with a buffer fund is sometimes not admissible because of the possibility of generating negative reserve during the leveling period. Of course, the presence in this variant of more important reserves makes this risk of non-admissibility less important.

4.2.4 Terminal Funding PAYG Method

This method is based on the principle of equivalence between, on the one hand, the contributions paid by the active population over a given period and, on the other hand, the lump sums used to fund the benefits (until death) of the generation retiring during that period.

Analytically, the contributions collected at time t (see the Pay-As-You-Go method) is

$$\pi_{\text{TF}}(t) \, \overline{S}(t) \sum_{x=x_0}^{x_r-1} L(x,t)$$

and for costs at t

$$\sum_{x=x_r}^{\omega} L(x, t+x-x_r) \, \overline{R}(t+x-x_r) \, v^{x-x_r}.$$

The application of the principle of equivalence therefore gives

$$\pi_{TF}(t) = \frac{\sum_{x=x_r}^{\omega} L(x, t+x-x_r) \overline{R}(t+x-x_r) v^{x-x_r}}{\overline{S}(t) \sum_{x=x_0}^{x_r-1} L(x,t)}.$$

In particular, assuming that the population is stationary ($L(x,t) = \chi(x)$) and that benefits are revaluated following a geometric progression with rate j, i.e.

$$\overline{R}(t+x-x_r) = \overline{R}(t)(1+j)^{x-x_r},$$

we obtain

$$\pi_{TF}(t) = \frac{\sum_{x=x_r}^{\omega} \chi(x) \cdot \overline{R}(t)(1+j)^{x-x_r} \cdot v^{x-x_r}}{\overline{S}(t) \sum_{x=x_0}^{x_0-1} \chi(x)}$$

$$= \frac{\chi(x_r) \cdot \overline{R}(t) \sum_{x=x_r}^{\omega} \frac{\chi(x)}{\chi(x_r)} \cdot (1+j)^{x-x_r} \cdot v^{x-x_r}}{\overline{S}(t) \sum_{x=x_0}^{x_r-1} \chi(x)}$$

$$= \frac{\chi(x_r)}{\sum_{x=x_0}^{x_r-1} \chi(x)} \cdot \delta(t) \cdot \sum_{x=x_r}^{\omega} p(x_r; x) \cdot (1+j)^{x-x_r} \cdot v^{x-x_r}$$

$$= \frac{\chi(x_r)}{\sum_{x=x_0}^{x_r-1} \chi(x)} \cdot \delta(t) \cdot \ddot{a}_{x_r},$$

where $\ddot{a}_{x_r}$ is the the annuity price at retirement age. In this case, the contribution rate can be written as the product of three factors:

- a *demographic* factor, giving the ratio between the number of people in the retiring class and the total number of working people,
- a *financial* factor given by the benefit ratio,
- a *life-time* factor, given by an annuity price.

In conclusion, let us summarize the main characteristics of the Terminal Funding method:

1. *constitution of reserves*: the scheme has permanent reserves equal to the retirees' reserve; a guarantee therefore exists but only for retirees,
2. *full inter-generational transfer*: the pension benefits of one generation are totally dependent on another generation,
3. *need for sustainability*: the method, which does not guarantee any future guarantee for workers, can only be conceived fairly in a philosophy of permanence of the pension system (only retirees benefit from a guarantee),
4. *inflation sensitivity*: the method depends on the future evolution of post-retirement benefits generally linked to inflation,

4.2 Classification of Funding Methods

5. *sensitivity to financial markets*: the method, generating strictly positive reserves, depends on the evolution of financial markets,
6. *sensitivity to demographic change*: the method depends on the demographic evolution of the population and especially on the balance between workers and retirees, but to a lesser extent than in the Pay-As-You-Go method.

4.2.5 Scaled Terminal Funding Method

As in the case of Pay-As-You-Go, the Terminal Funding method can be spread over several years, giving rise to the *Scaled Terminal Funding method*. In this case, the actuarial equivalence can be written as

Discounted value of workers contributions received between t and $t+T$

$$=$$

Discounted value of lump sums needed to pay all benefits

of generations retiring between t and $t+T$.

Analytically, the current value of contributions in $[t, t+T]$ is

$$\pi_{SC}(t) \cdot \sum_{s=0}^{T} \overline{S}(t+s) \left(\sum_{x=x_0}^{x_r-1} L(x, t+s) \right) v^s$$

and the discounted value of the underlying lump sum in $[t, t+T]$ is

$$\sum_{s=0}^{T} \sum_{x=x_r}^{\omega} L(x, t+s+x-x_r) \overline{R}(t+s+x-x_r) \cdot v^{s+x-x_r}.$$

The application of the principle of equivalence gives

$$\pi_{SC}(t) = \frac{\sum_{s=0}^{T} \sum_{x=x_r}^{\omega} L(x, t+s+x-x_r) \overline{R}(t+s+x-x_r) v^{s+x-x_r}}{\sum_{s=0}^{T} \overline{S}(t+s) \sum_{x=x_0}^{x_r-1} L(x, t+s) \cdot v^s}.$$

4.2.6 Fully Funded Methods

During their working lives, each individual finances his/her own retirement. In this case, there is an equivalence between the discounted value of contributions paid during the career and the discounted value of benefits paid after retirement.

Analytically, for the group starting to work at t, the discounted value of contributions is

$$\sum_{x=x_0}^{x_r-1} L(x; t+x-x_0) \cdot \overline{S}(t+x-x_0) \cdot \pi_{FF}(x) \cdot v^{x-x_0},$$

where $\pi_{FF}(x)$ is the contribution rate to be applied at age x. For the discounted value of costs we obtain

$$\sum_{x=x_r}^{\omega} L(x; t+x-x_0) \cdot \overline{R}(t+x-x_0) \cdot v^{x-x_0}.$$

The application of the principle of equivalence gives

$$\sum_{x=x_0}^{x_r-1} L(x; t+x-x_0) \cdot \overline{S}(t+x-x_0) \cdot \pi_{FF}(x) \cdot v^{x-x_0}$$

$$= \sum_{x=x_r}^{\omega} L(x, t+x-x_0) \cdot \overline{R}(t+x-x_0) \cdot v^{x-x_0}. \qquad (4.8)$$

A Fully Funded method is thus defined by a vector of successive contribution rates

$$\{\pi_{FF}(x_0), \pi_{FF}(x_0+1), \cdots, \pi_{FF}(x_r-1)\}$$

satisfying Eq. (4.8). There are therefore an infinite number of Fully Funded methods.

One can consider the particular case where a constant contribution rate is applied throughout the career: if

$$\pi_{FF}(x_0) = \pi_{FF}(x_0+1) = \cdots = \pi_{FF}(x_r-1) = \pi_{FF},$$

then we have

$$\pi_{FF} = \frac{\sum_{x=x_r}^{\omega} L(x; t+x-x_0) \overline{R}(t+x-x_0) v^{x-x_0}}{\sum_{x=x_0}^{r-1} L(x, t+x-x_0) \overline{S}(t+x-x_0) v^{x-x_0}}.$$

If we also assume that the population is stationary, that benefits are revaluated following a geometric progression with rate j and that salaries are revaluated

4.2 Classification of Funding Methods

following a geometric progression with rate g (benefits and salaries can be indexed at different rates), we obtain

$$\pi_{FF} = \frac{\sum_{x=x_r}^{\omega} \chi(x) \cdot \overline{R}(t + x_r - x_0) \cdot (1+j)^{x-x_r} \cdot v^{x-x_0}}{\sum_{x=x_0}^{x_r-1} \chi(x) \cdot \overline{S}(t) \cdot (1+g)^{x-x_0} \cdot v^{x-x_0}}$$

$$= \frac{\overline{R}(t + x_r - x_0)}{\overline{S}(t)} \frac{\left(\frac{\chi(x_r)}{\chi(x_0)} v^{x_r-x_0}\right) \sum_{x=x_r}^{\omega} \frac{\chi(x)}{\chi(x_r)} (1+j)^{x-x_r} \cdot v^{x-x_r}}{\sum_{x=x_0}^{x_r-1} \frac{\chi(x)}{\chi(x_0)} (1+g)^{x-x_0} \cdot v^{x-x_0}}$$

$$= \frac{\overline{R}(t + x_r - x_0)}{\overline{S}(t)} \cdot {}_{x_r-x_0}E_{x_0} \cdot \frac{\ddot{a}_{x_r}^{(j)}}{\ddot{a}_{x_0;x_r-x_0}^{(g)}}.$$

In this case, the contribution rate is the product of the three following factors:

- a *benefit ratio* $\frac{\overline{R}(t+x_r-x_0)}{\overline{S}(t)}$, representing the ratio between the first benefit at retirement age and the first salary,
- the price of a *pure endowment* between the age of entry into service and the retirement age,
- the ratio between *two annuity prices*.

To conclude, let us summarize the main characteristics of the Fully Funded methods:

1. *constitution of reserves*: the scheme has permanent reserves equal to the retirees' reserve. In addition, the workers have a reserve corresponding to their own benefits,
2. *no inter-generational transfer*: there is no transfer between generations, as each generation funds its own retirement,
3. *need for permanence*: the method can be applied whether or not there is a guarantee of the pension scheme's permanence,
4. *sensitivity to inflation*: the method depends on the future evolution of salaries and benefits after retirement,
5. *sensitivity to financial markets*: the method leading to the creation of large reserves, it depends strongly on the evolution of financial markets,
6. *insensitivity to demographic change*: since each generation is self-supporting, there is no dependence on the evolution of the demographic population of successive generations.

   However, demography affects the method through the longevity risk associated with the relevance of the life tables used in relation to the real evolution of the population. As already observed in Chap. 2, it can be argued that Fully Funded methods are insensitive to population growth risk but highly dependent on longevity risk.

4.2.7 Numerical Example

We propose to compare numerically the different funding methods, based on the following simple assumptions:

1. An initial population of 600 people with a population function given by

$$L(x, 0) = \begin{cases} 10 & \text{if } 20 \leq x < 30 \\ 20 & \text{if } 30 \leq x < 50 \\ 10 & \text{if } 50 \leq x < 60 \\ 0 & \text{if } x \geq 60. \end{cases}$$

2. The only cause of exit from the population is death, assumed to occur uniformly at age 75 (no mortality before that age).
3. At the entry level, two cases are considered: on the one hand, a closed population without replacements; on the other hand, an open population in which each individual retiring at age 65 is replaced by a 20-year worker ($E(20, t) = L(65, t)$).
4. We work with a zero interest rate.
5. A pension scheme is introduced at $t = 0$, granting a benefit equal to half of the final salary. Salaries are assumed to be uniform over the population. Neither salaries nor benefits are revaluated.

We will compare three funding methods: Pay-As-You-Go, Terminal Funding and Fully Funded methods, successively for a closed and an open population.

4.2.7.1 Closed Population

Pay-As-You-Go Method

Given the assumptions, the contribution rate at any time t is given by:

$$\pi_{PA}(t) = 0.5 \frac{\text{number of retirees}}{\text{number of workers}},$$

whence

for $t < 5$,

$$\pi_{PA}(t) = 0 \quad \text{(no retiree)},$$

for $t = 6$,

$$\pi_{PA}(6) = 0.5 \frac{10}{590} = 0.0085 \quad \text{(first generation of retirees)},$$

4.2 Classification of Funding Methods

for $t = 7$,
$$\pi_{PA}(7) = 0.5 \frac{20}{580} = 0.017,$$

for $t = 15$,
$$\pi_{PA}(15) = 0.5 \frac{100}{500} = 0.1,$$

for $t = 16$,
$$\pi_{PA}(16) = 0.5 \frac{9 \times 10 + 1 \times 20}{480} = 0.11,$$

for $t = 25$,
$$\pi_{PA}(25) = 0.5 \frac{200}{300} = 0.33,$$

for $t = 35$,
$$\pi_{PA}(35) = 0.5 \frac{200}{100} = 1,$$

for $t > 45$, there are no longer any workers and the scheme must stop. No more benefits can be paid, neither to current workers nor to generations already retired.

Terminal Funding Method

In this case, the annuity price at age 65, $\ddot{a}_{65}$ is equal to 10, and the contribution rate is given by

$$\pi_{TF}(t) = 0.5 \cdot \ddot{a}_{65} \cdot \frac{\text{number of individuals becoming retirees}}{\text{number of workers}},$$

whence

for $t < 5$,
$$\pi_{TF}(t) = 0,$$

for $t = 6$,
$$\pi_{TF}(6) = 0.5 \cdot 10 \cdot \frac{10}{590} = 0.085,$$

for $t = 7$,
$$\pi_{TF}(7) = 0.5 \cdot 10 \cdot \frac{10}{580} = 0.086,$$

for $t = 15$,
$$\pi_{TF}(15) = 0.5 \cdot 10 \cdot \frac{10}{500} = 0.1,$$

for $t = 16$,
$$\pi_{TF}(16) = 0.5 \cdot 10 \cdot \frac{20}{480} = 0.21,$$

for $t = 25$,
$$\pi_{TF}(25) = 0.5 \cdot 10 \cdot \frac{20}{300} = 0.33,$$

for $t = 35$,
$$\pi_{TF}(35) = 0.5 \cdot 10 \cdot \frac{20}{100} = 1,$$

for $t > 45$, there are no longer any workers and the funding stops. The workers present at that time will never receive benefits. However, generations who have already retired will continue to receive their benefits (as they have already been fully funded).

Fully Funded Method

Each age group will pay a different contribution rate, assumed to be constant throughout their career. Denoting by $\pi_{FF}(x)$ the contribution rate for the initial age group x at $t = 0$ (x between 20 and 60), we get, taking into account the chosen parameters:

$$\pi_{FF}(x) = 0.5 \cdot \frac{\ddot{a}_{65}}{65 - x},$$

whence

for the group being 20 years old at $t = 0$,

$$\pi_{FF}(20) = \frac{0.5 \times 10}{45} = 0.11,$$

4.2 Classification of Funding Methods

for the group being 30 years old at $t = 0$,

$$\pi_{FF}(30) = \frac{0.5 \times 10}{35} = 0.14,$$

for the group being 50 years old at $t = 0$,

$$\pi_{FF}(50) = \frac{0.5 \times 10}{15} = 0.33,$$

for the group being 60 years old at $t = 0$,

$$\pi_{FF}(60) = \frac{0.5 \times 10}{5} = 1.$$

This simple example illustrates in a closed population the phenomena of *first* and *last generation* and the fundamental antagonism of treatment between Pay-As-You-Go and Fully Funded methods.

Let us consider the first generation of retirees, i.e. the group initially aged 60. In the Pay-As-You-Go method, this group does not pay any contributions, and receives a full benefit until death. In the Fully Funded method, this group has to fund all the benefits to be received within 5 years, and has a maximum contribution rate. The Pay-As-You-Go method therefore grants the so-called "first generation gift".

In contrast, let us consider the last generation of the population, i.e. the group initially aged 20 years. In the Pay-As-You-Go method, this group has had to pay increasing contributions (for 35 years) in order to receive no benefits when they retire. In the Fully Funded method, this group can spread its expense over 45 years and has a minimum contribution rate; its pension benefits are insured. The "first generation gift" in the Pay-As-You-Go method is therefore transformed into a "last generation nightmare" in a closed population.

4.2.7.2 Open Population

Pay-As-You-Go Method

Taking into account the replacement assumptions, the working population maintains a constant size of 600 workers. The contribution rate therefore becomes

$$\pi_R(t) = 0.5 \cdot \frac{\text{number of retirees}}{600},$$

whence

for $t < 5$,
$$\pi_{PA}(t) = 0,$$

for $t = 6$,
$$\pi_{PA}(6) = 0.5 \cdot \frac{10}{600} = 0.0083,$$

for $t = 7$,
$$\pi_{PA}(7) = 0.5 \cdot \frac{20}{600} = 0.016,$$

for $t = 15$,
$$\pi_{PA}(15) = 0.5 \cdot \frac{100}{600} = 0.083,$$

for $t = 16$,
$$\pi_{PA}(16) = 0.5 \cdot \frac{9 \times 10 + 1 \times 20}{600} = 0.092,$$

for $t = 25$,
$$\pi_{PA}(25) = 0.5 \cdot \frac{200}{600} = 0.166,$$

for $t = 35$,
$$\pi_{PA}(35) = 0.5 \cdot \frac{200}{600} = 0.166.$$

The system can continue indefinitely.

Terminal Funding Method

The conclusion is the same as in the Pay-As-You-Go method. All denominators appearing in the contribution rate are replaced by the constant number of workers (i.e. 600), and the system can continue.

Fully Funded Method

For generations present at $t = 0$, the contribution rate is the same as for the closed population (no solidarity between generations).

4.3 Funded or Pay-As-You-Go Scheme: The Samuelson–Aaron Rule

The purpose of this section is to generalize in a multi-period setting the two-period model seen in Sect. 3.5.3 and to find a theoretical equivalence relationship between the Pay-As-You-Go and Fully Funded methods: can we find conditions implying the supremacy of one of the two funding methods?

4.3.1 First and Last Generations

We consider here a system in stationary state: the pension system is supposed to exist for a long period of time and is not doomed to disappear soon. Otherwise, we would face the so-called first and last generation phenomena, and the comparison would be biased.

Indeed, in this case, the initial group of retirees, even though they have never contributed, will be able to receive a full pension directly under Pay-As-You-Go: this is the first generation gift. On the other hand, in the Fully Funded scheme, if they have not contributed, these "initial" retirees will not be entitled to any benefits.

Symmetrically, the last generation phenomenon occurs when the financing of the plan comes to a definitive end; in this case, the working people who have previously contributed will no longer receive any benefits under the Pay-As-You-Go system, whereas under the Fully Funded scheme, they will receive a pension corresponding to the sums saved previously.

In these two extreme situations, the comparison between pure Pay-As-You-Go and Fully Funded schemes becomes obvious.

4.3.2 Funded and Pay-As-You-Go Schemes in Stationary Phase

Let us therefore consider a system in a stationary state in order to avoid these extreme cases. We therefore assume the population to be open. The assumptions used for the comparison will be as follows:

1. demographic assumptions: the population, described through a discrete model, is stable; the population function is of the form (cf. Eq. (3.8)):

$$L(x, t) = A \cdot e^{\rho t} e^{-\rho(x-x_0)} p(x_0, x),$$

where x_0 is the age of entry into active life. We define x_r as the age of retirement,

2. macroeconomic assumptions: we define

$$\sigma = \text{continuous growth rate of the salaries and of the pension benefits,}$$
$$\gamma = \text{continuous interest rate,}$$
$$\overline{S}(t) = \text{average salary at time } t.$$

3. DC pension scheme assumptions: the contribution rate, denoted π, is constant. We want to compare the benefits obtained in the two methods. We define

$$\overline{R}_{\text{PA}}(t) = \text{average benefit obtained in the Pay-As-You-Go method,}$$
$$\overline{R}_{\text{FF}}(t) = \text{average benefit obtained in the Fully Funded method.}$$

In the Pay-As-You-Go method, we only have to match incomes and outcomes at time t (vertical approach). The incomes are equal to

$$I_t = \pi \cdot \overline{S}(t) \cdot \sum_{x=x_0}^{x_r-1} L(x,t) = \pi \cdot A \cdot S(t) \cdot e^{\rho t} \sum_{x=x_0}^{x_r-1} e^{-\rho(x-x_0)} p(x_0; x),$$

while the outcomes are equal to

$$II_t = \overline{R}_{\text{PA}}(t) \cdot \sum_{x=x_r}^{\omega} L(x,t) = A \cdot \overline{R}_{\text{PA}}(t) \cdot e^{\rho t} \sum_{x=x_r}^{\omega} e^{-\rho(x-x_0)} p(x_0; x).$$

Writing that these two quantities are equal yields

$$\overline{R}_{\text{PA}}(t) = \pi \cdot \overline{S}(t) \cdot \frac{\sum_{x=x_0}^{x_r-1} e^{-\rho(x-x_0)} p(x_0; x)}{\sum_{x=x_r}^{\omega} e^{-\rho(x-x_0)} p(x_0; x)}. \qquad (4.9)$$

In the Fully Funded method, the discounted value of contributions must be equal to the discounted value of the benefits (cohort entering working life at time t) over the entire duration of the career assumed to begin at time t. The incomes are equal to

$$I_t = L(x_0, t) \sum_{x=x_0}^{x_r-1} \pi \cdot S(t) \cdot e^{\sigma(x-x_0)} \cdot e^{-\gamma(x-x_0)} \cdot p(x_0; x)$$

$$= L(x_0, t) \pi \cdot S(t) \sum_{x=x_0}^{x_r-1} e^{-(\gamma-\sigma)(x-x_0)} p(x_0; x),$$

4.3 Funded or Pay-As-You-Go Scheme: The Samuelson–Aaron Rule

while the outcomes are equal to

$$\text{II}_t = L(x_0, t) \cdot \overline{R}_{\text{FF}}(t) \sum_{x=x_r}^{\omega} e^{\sigma(x-x_0)} \cdot e^{-\gamma(x-x_0)} p(x_0; x).$$

By equivalence between these two expressions we get

$$\overline{R}_{\text{FF}}(t) = \pi \cdot S(t) \cdot \frac{\sum_{x=x_0}^{x_r-1} e^{-(\gamma-\sigma)(x-x_0)} p(x_0; x)}{\sum_{x=x_r}^{\omega} e^{-(\gamma-\sigma)(x-x_0)} p(x_0; x)}. \tag{4.10}$$

The comparison of Eqs. (4.9) and (4.10) shows that the benefits obtained in both methods are of the same form. This can be made crystal clear by introducing the function

$$\varphi(\zeta) = \frac{\sum_{x=x_0}^{x_r-1} e^{-\zeta(x-x_0)} p(x_0, x)}{\sum_{x=x_r}^{\omega} e^{-\zeta(x-x_0)} p(x_0, x)},$$

so that Eqs. (4.9) and (4.10) become

$$\overline{R}_{\text{PA}}(t) = \pi \cdot s(t) \cdot \varphi(\rho) \quad \text{and} \quad \overline{R}_{\text{FF}}(t) = \pi \cdot s(t) \cdot \varphi(\gamma - \sigma).$$

To compare $\overline{R}_{\text{PA}}(t)$ and $\overline{R}_{\text{FF}}(t)$, it is therefore necessary to study the behavior of the function φ, which is the goal of the following lemma.

Lemma 4.1 *The function is φ monotonous and non-decreasing.*

Proof Since φ is clearly differentiable, it is sufficient to show that its derivative is non-negative.

To reduce the amount of notation, let us define

$$\Theta_a = \sum_{x=x_0}^{x_r-1} e^{-\zeta(x-x_0)} p(x_0, x),$$

$$\Theta_r = \sum_{x=x_r}^{\omega} e^{-\zeta(x-x_0)} p(x_0, x).$$

We then compute

$$\frac{\partial \varphi}{\partial \zeta} = \frac{1}{\Theta_r^2} \left(\Theta_r \cdot \left[\sum_{x=x_0}^{x_r-1} -(x-x_0) e^{-\zeta(x-x_0)} p(x_0, x) \right] \right.$$

$$\left. - \Theta_a \cdot \left[\sum_{x=x_r}^{\omega} -(x-x_0) e^{-\zeta(x-x_0)} p(x_0, x) \right] \right).$$

Two inequalities lead to the conclusion:

$$\sum_{x=x_r}^{\omega}(x-x_0)\,e^{-\zeta(x-x_0)}p(x_0;x) \geq (x_r-x_0)\,\Theta_r$$

together with

$$\sum_{x=x_0}^{x_r-1}(x-x_0)\,e^{-\zeta(x-x_0)}p(x_0;x) \leq (x_r-x_0)\,\Theta_a$$

show that

$$\frac{\partial\varphi}{\partial\zeta} \geq \frac{(x_r-x_0)\,\Theta_a\cdot\Theta_r - (x_r-x_0)\,\Theta_r\Theta_a}{\Theta_r^2} = 0.$$

□

This lemma provides the following corollary, which generalizes the comparison Eq. (3.12) seen in the two-period model.

Corollary 4.1

$$\overline{R}_{\text{PA}}(t) \geq \overline{R}_{\text{FF}}(t) \Leftrightarrow \rho \geq \gamma - \sigma$$

The Pay-As-You-Go method provides larger benefits than the Fully Funded method if the population growth rate is larger than the difference between the interest rate and the salary growth rate. In this case, the first generation gift is "propagated" from generation to generation: thanks to the demographic dynamics, each generation receives more in benefits than the capitalized value of its contributions. This is the Samuelson–Aaron paradox or the Social Security paradox: the first generation gift in PAYG is never paid and on the contrary each generation receives an "inter-generational gift".

In contrast, in periods of high market interest rates (i.e. $\rho < \gamma - \sigma$), the Fully Funded method should be preferred. Another way to express this comparison is based on the alternative form of Corollary 4.1: $\rho + \sigma \geq \gamma$. The return of a fully funded scheme is the financial return γ; the return of Pay-As-You-Go is the sum of the growth of salary σ and growth of population ρ. This sum represents the growth rate of the total payroll and can be therefore seen as the mean return of a Pay-As-You-Go scheme.

The behavior of these three parameters can be examined backwards to see what the funding choice should have been. We can say that in the 1950s and 1960s, our countries experienced a significant population growth rate (cf. the "Baby boom"), interest rates barely followed inflation and salaries rose steadily above inflation. So we were clearly in the $\rho > \gamma - \sigma$ scenario and the Pay-As-You-Go method was preferable.

4.3 Funded or Pay-As-You-Go Scheme: The Samuelson–Aaron Rule

On the contrary, from 1970 to 2000, the population growth rate was almost equal to zero, and the interest rate significantly exceeded the inflation rate, which had become very low. So we were in the $\rho < \gamma - \sigma$ scenario and the Fully Funded method became preferable.

4.3.3 Terminal Funding Schemes in Stationary Phase

In addition to the Pay-As-You-Go and Fully Funded methods, we can also consider the Terminal Funding method in the above model. We obtain in this case

$$I_t = \pi \cdot A \cdot S(t) e^{\rho t} \sum_{x=x_0}^{x_r-1} e^{-\rho(x-x_0)} p(x_0; x)$$

for the incomes. For each retiree at t, we need to have a lump sum equal to

$$\sum_{x=x_r}^{\omega} \overline{R}_{\text{TF}}(t) \cdot e^{-(\gamma-\sigma)(x-x_r)} p(x_r; x),$$

where $\overline{R}_{TF}(t)$ is the average benefit in Terminal Funding that can be provided as lump sum. This lump sum must be provided for the entire cohort reaching retirement at time t, whose size is given by

$$L(x_r, t) = A \cdot e^{\rho t} e^{-\rho(x_r-x_0)} \cdot p(x_0; x_r).$$

We therefore obtain, for the outcomes,

$$\text{II}_t = A \cdot e^{\rho t} e^{-\rho(x_r-x_0)} p(x_0; x_r) \overline{R}_{\text{TF}}(t) \sum_{x=x_r}^{\omega} e^{-(\gamma-\sigma)(x-x_r)} p(x_r; x)$$

$$= A \cdot e^{\rho t} e^{-\rho(x_r-x_0)} \overline{R}_{\text{TF}}(t) \sum_{x=x_r}^{\omega} e^{-(\gamma-\sigma)(x-x_0)} p(x_0; x).$$

Writing that these two quantities are equal yields

$$\overline{R}_{\text{TF}}(t) = \pi \cdot S(t) = \frac{\sum_{x=x_0}^{x_r-1} e^{-\rho(x-x_0)} p(x_0; x)}{\sum_{x=x_r}^{\omega} e^{-[\rho(x_r-x_0)+(\gamma-\sigma)(x-x_r)]} p(x_0, x)}, \tag{4.11}$$

which we can compare with Eqs. (4.9) and (4.10).

In particular, if $\rho = \gamma - \sigma$, the three methods are equivalent:

$$\overline{R}_{\text{TF}}(t) = \overline{R}_{\text{PA}}(t) = \overline{R}_{\text{FF}}(t).$$

Formula (4.11) illustrates the intermediate nature of the Terminal Funding method, between Pay-As-You-Go and Fully Funded methods.

Part II
Pay-As-You-Go Pension Schemes

In this second part, Pay-As-You-Go techniques are studied in the context of financing social security schemes. Given the almost generalized use of Pay-As-You-Go in the financing of social security schemes, the latter are analyzed in detail. In particular, the difficulties generated by these schemes are highlighted. In order to respond to the challenges posed to these first pillars, two alternative techniques are considered: the constitution of a balancing fund through so-called leveling techniques on the one hand, and the variation of Pay-As-You-Go into Defined Contribution on the other. In particular, the techniques of notional accounts (NDC) and points are developed.

Chapter 5
Pay-As-You-Go Social Security Schemes

Abstract The Pay-As-You-Go method is naturally associated in people's minds with social security pension schemes. Indeed, even if some countries, such as Chile, have moved to fully funded methods, the vast majority of public systems have adopted the Pay-As-You-Go philosophy. On the other hand, and symmetrically, the fully funded system is generally required for supplementary schemes. It is therefore useful to briefly describe the specific aspects of managing first pillar schemes following the pas-as-you-go method.

5.1 General Principles

Virtually all social security schemes are characterized by two elements, one relating to the choice of benefits and the other to the method of funding. In terms of benefits, most of these schemes adopt a Defined Benefit philosophy. This choice stems from the desire for transparency of the public system towards members. Since the first pillar pension constitutes a first basic level to be completed (or even sometimes a simple safety net), it is important for everyone to know as precisely as possible the extent of the benefits to be granted.

In terms of funding, the Pay-As-You-Go method is almost universally used. Many arguments can be put forward to justify this choice: the permanence of the system, the large population, the inter-generational solidarity, the lack of dependence on inflation, and the fast adaptability.

In contrast, a complete transition to the fully funded method would often require provisions too large in terms of savings capacity.

On the other hand, while the Samuelson–Aaron rule may sometimes lead to a temporary preference for fully funded schemes, the transition from a Pay-As-You-Go system to a fully funded system is particularly delicate if we want to avoid any discontinuity in the payment of benefits. In this case, the generation of workers may have to pay twice: once for itself in the form of savings, and once for the generation of retirees at the time of transition, who have never had the opportunity to save.

Of course, the reverse transition from a fully funded to a PAYG system does not pose a problem. On the contrary, it suddenly frees up reserves made superfluous in

© The Author(s), under exclusive license to Springer Nature Switzerland AG 2025
P. Devolder, S. de Valeriola, *Actuarial Pension Funding Theory*, Springer Actuarial Textbooks, https://doi.org/10.1007/978-3-031-85268-8_5

the new method... This shift has been made in some European countries after the Second World War.

The challenge in building a first pillar as a Defined Benefit scheme is therefore generally less a choice of a funding method than a control of benefits in order to avoid an excessive dependence of the overall amount of pensions on this first pillar alone. Indeed, the particular nature of the three pension pillars leads quite naturally to the following combination:

- First pillar: social security—Pay-As-You-Go method,
- Second pillar: supplementary occupational schemes—collective or individual fully funded method,
- Third pillar: individual savings—individual fully funded method.

A balanced place for each of these pillars makes it easy to adapt to the performance conditions dictated by the Samuelson–Aaron rule. This strategy of diversifying retirement methods is analogous with the principles of sound portfolio theory diversification. This approach "multi-pillar" and "multi-funding" can also be enriched by complementary components, such as a "first pillar bis" using the fully funded method and Defined Contributions (see Sect. 5.5).

While the Pay-As-You-Go philosophy is therefore the natural framework for first pillar schemes (or at least a first tier of them), its pure and lasting application can lead to a particularly volatile and unstable funding structure.

Indeed, in a Defined Benefit plan, the contribution rate is irrevocably fixed by the relationship:

$$\text{contribution rate} = \frac{\text{pension costs}}{\text{salaries}}.$$

This rate may vary more or less significantly from one year to the next.

In social security schemes for employed workers, three sources of funding are often used:

- a *personal contribution*, paid by workers and generally deducted from their salary. The contribution is expressed as a rate applied to the (possibly capped) salary,
- an *employer contribution*, paid by the employer, in addition to salary. It is also generally expressed in the form of a rate applied to capped or uncapped salary,
- an *intervention* complementary to the State budget to supplement the above contributions (whose rates are generally fixed) to reach the required level.

The essentially fluctuating nature of the pure Pay-As-You-Go contribution rate may lead to the consideration of smoothing its effects. One possible answer is the Scaled Premium method (Sect. 4.2.2), which consists in applying the Pay-As-You-Go method over several years.

Another similar technique aims to project the Pay-As-You-Go method directly over several years, and then apply a leveling. These leveling and equalization techniques, implying the creation of a buffer fund (and therefore at the semantic level a partial transition to funded methods), will be presented in Sect. 5.6, below.

Finally, it should be noted that while, as mentioned above, the vast majority of first pillar Pay-As-You-Go schemes are in Defined Benefit, it is also possible to consider Pay-As-You-Go schemes in Defined Contribution or hybrid between DB and DC. Chapters 6–8 will be devoted to such schemes.

5.2 Internal Rate of Return of a Pay-As-You-Go Scheme

It is not easy for the affiliate of a Pay-As-You-Go scheme to assess individual performance of the system. One possible way is to compare the contributions that will be paid during working life with the benefits that will be paid from retirement age.

As an example, let us consider a Pay-As-You-Go scheme planning a fixed contribution as a percentage of salary from the age of 25 and granting an indexed retirement pension corresponding to a fixed percentage of the last salary from the age of 65. Pensions are assumed to be revaluated as salaries.

What is the internal rate of return on this operation? Let us compare the discounted value at retirement age (using the desired rate of return) of the contributions and the benefits. In order to simplify the approach at the demographic level, we will ignore pre-retirement mortality and assimilate life annuity after retirement to some financial annuity whose duration corresponds to life expectancy at age 65, set for illustration at 15 years.

So we have, for a unitary final salary at age 65:

- for contributions, the accumulated value of contributions at retirement age:

$$C_1 = \pi \sum_{x=25}^{64} \frac{1}{(1+g)^{65-x}} (1+i)^{65-x},$$

where i is the rate of return to compute, π is the contribution rate and g is the annual rate of salary growth.
- for benefits, the discounted value of benefits at retirement age:

$$C_2 = \delta \sum_{x=65}^{79} (1+g)^{x-65} \frac{1}{(1+i)^{x-65}},$$

Table 5.1 Examples of benefit ratios and Pay-As-You-Go rates of return

Rate of return i	Benefit ratio δ
3%	31%
4%	40%
5%	52%
6%	69%
6,05%	70%

where δ is the benefit ratio. One can, for different values of i, compute δ such that the quantities C_1 and C_2 are equal. Assuming, for example, a contribution rate of 15% and a revaluation rate of 4%, we obtain

$$0,15 \sum_{x=25}^{64} \frac{1}{(1,04)^{65-x}}(1+i)^{65-x} = \delta \cdot \sum_{x=65}^{79}(1,04)^{65-x}\frac{1}{(1+i)^{65-x}}$$

or, with $k = \frac{1+i}{1.04}$,

$$\delta = \begin{cases} 0,15 \cdot \frac{k^{40}-1}{1-\left(\frac{1}{k}\right)^{15}} & \text{for } i \neq 0,04, \\ 0,15 \cdot \frac{40}{15} = 0,40 & \text{for } i = 0,04. \end{cases}$$

Table 5.1 gives the equilibrium benefit ratio for different values of the rate of return. Thus a benefit ratio of 70% of the last salary would correspond to an internal rate of return of 6.05%.

In addition to the affiliate's judgment on the profitability of his/her pension operation, the return obtained also makes it possible to judge the viability of the Pay-As-You-Go plan, thanks to the Samuelson–Aaron rule and the equivalence relationships of Corollary 4.1, under a hypothesis of demographic stationarity.

The Pay-As-You-Go scheme will be in equilibrium if the rate of population growth is such that

$$\rho \geq \gamma - \sigma \quad \text{or} \quad e^\rho \geq e^{\gamma-\sigma} = \frac{e^\gamma}{e^\sigma} = \frac{1+i}{1,04},$$

i.e., for a replacement rate of 70%, a minimum annual population increase of 1.97%.

5.3 A Macroeconomic Indicator

While the traditional actuarial approach naturally leads to expressing pension expenditures as a proportion of the total contributory salaries, another interesting ratio of a more economic nature is the ratio of pension expenditure to Gross

5.4 Parameters of a Social Security Scheme

Domestic Product (GDP), which measures the capacity of a country's economy to bear the pension costs of its first pillar.

By assimilating firstly, as before, on the one hand, the active population to the workforce between the entry age x_0 and the retirement age x_r and, on the other hand, the retired population to the population beyond the retirement age x_r, we obtain

- for pension expenditure at t

$$\overline{R}(t) \cdot \sum_{x=x_r}^{\omega} L(x,t),$$

where $\overline{R}(t)$ is the average pension benefit at time t;
- for the Gross Domestic Product at t

$$\text{pr}(t) \sum_{x=x_0}^{x_r-1} L(x,t),$$

where $\text{pr}(t)$ is the average productivity at time t;
- for the GDP ratio

$$\tau_{\text{GDP}}(t) = \frac{\overline{R}(t) \cdot \sum_{x=x_r}^{\omega} L(x,t)}{\text{pr}(t) \cdot \sum_{x=x_0}^{x_r-1} L(x,t)} = D(t) \cdot \frac{\overline{R}(t)}{\text{pr}(t)}.$$

The evolution of this ratio therefore depends on two phenomena: the demographic dynamics of the population (expressed by its dependence ratio), and the revaluation of the average pension in relation to average productivity growth.

If, in a more general way, an unemployment rate $u(t)$ is introduced—a part of the population between the ages x_0 and x_r is assumed to be inactive—we obtain

$$\tau_{\text{GDP}}(t) = \frac{\overline{R}(t) \cdot \sum_{x=x_r}^{\omega} L(x,t)}{\text{pr}(t) \cdot (1-u(t)) \sum_{x=x_0}^{x_r-1} L(x,t)} = D(t) \cdot \frac{1}{(1-u(t))} \frac{\overline{R}(t)}{\text{pr}(t)}.$$

5.4 Parameters of a Social Security Scheme

A Social Security pension scheme generally includes the definition of the following elements:

- categories of beneficiaries,
- the normal retirement age,
- the calculation of the benefit formula.

5.4.1 Categories of Beneficiaries

A Social Security scheme may cover the entire population. It may also cover only one category of workers.

For instance we could have different social security arrangements for employees, civil servants and the self-employed.

In some countries special schemes exist for executives. Another possible distinction within salaried employees is that between white-collar and blue-collar workers. There are also often special schemes for seafarers, miners, ...

Special rules must be made in the case of mixed careers, i.e. workers moving from one category to another during their career.

Irrespective of the occupational distinction, other discriminatory criteria may be met, either to exclude certain individuals from entitlement to benefits or to define different levels of benefits. Here, for example, we might think of:

- nationality conditions: some schemes exclude workers who are not nationals of the country,
- gender distinctions: Social Security schemes could provide different benefits for men and women, both in terms of retirement age, level of benefits or definition of derived rights. In Europe, within the framework of European Community law, any distinction according to sex in a Social Security scheme is now prohibited and all schemes must equalize conditions, subject to possible transitional provisions.

Finally, some workers may voluntarily (on an individual basis) exclude themselves from schemes to which they would be entitled under the regulations. This is for example the "Contracting out" scheme: in return for a reduction in his Social Security contributions, the worker may waive part of the benefits provided by Social Security, compensating for this lack under the second pillar or even the third pillar. In contrast to the classic paradigm of accumulation of the different pension pillars, we observe here a philosophy of substitution between pillars.

5.4.2 Normal Retirement Age

The normal retirement age is the reference age from which benefits can be paid without special early retirement penalties. Many schemes do provide for the possibility of early retirement before this age but with reduced benefits.

Historically, many schemes provided for different retirement ages for men and women. Under European legislation, these distinctions are forbidden.

In general, the normal retirement age in many European countries was historically equal to 65. There is a trend in many countries to delay the pension beyond the age of 65.

5.5 Public Pension Reforms

Many public pension schemes are facing major budgetary challenges in the context of aging populations. Indeed, these systems contain the germ of a major "social tension" that appears in periods of demographic instability: the impossible dream of combining Defined Benefits and Defined Contributions. Affiliates have received a contractual promise on a level of benefits that they hardly want to give up, but at the same time, as the funding is mainly ensured by contributions deducted from their income, they refuse any increase in these contributions.

The demographic outlook for the coming decades will only serve to accentuate this impossible wide gap. Faced with this major challenge and with a view to avoiding an irremediable shipwreck, various solutions have been developed by the public authorities, some of which can be combined.

Below are a few of these avenues for cleaning up the pension system. Some underlying actuarial methods will be explained in the following chapters.

5.5.1 Alternative Sources of Funding

In order to maintain both the level of benefits and the level of contributions on salaries, the State supplements each year to the extent due, either through taxation or by using alternative sources of funding. This type of technique is conceivable and has been implemented when the cost drift is slight. It is of course much more tricky to use when the size of the deficits become huge.

5.5.2 Creation of a Buffer Fund

In addition to the Pay-As-You-Go funding that is maintained, a fully funded mechanism is set up through a balanced fund, which is fed at the public level and allows, when pension expenditure exceeds contributions, the imbalance to be funded. The scheme is therefore only a very partial fully funded one: its reserves are in principle expected to be emptied rapidly.

This technique, which undoubtedly has its budgetary virtues and makes it possible, if necessary, to recycle public war treasures in an intelligent manner, is once again not in a position to meet the scale of the challenge on its own. This idea has already been introduced in Sect. 4.2.2 and will developed in Sect. 5.6.

5.5.3 Parametric Reforms

Parametric reforms aim to tackle the source of the benefit-contribution imbalance while maintaining the architecture of the scheme. The principle of a Pay-As-You-Go Defined Benefit plan is retained, but the parameters are changed to restore (or come closer to) equilibrium.

Examples of such adaptations are

- increasing the contribution rate,
- increasing the legal retirement age,
- tightening the conditions for early retirement,
- decreasing the caps on the salaries used to calculate benefits,
- lowering the benefit rates,
- shifting from a system based on final salaries to a system based on average salaries (e.g. over the entire career),
- revising the revaluation of pension benefits after retirement age.

If this parametric approach could theoretically stabilize the system, its concrete implementation often proves to be politically very delicate. It requires time and consistency in decisions. Of course, it comes up against public opinion by highlighting points of frustration in particular. In practice, the adjustments are then largely insufficient and sometimes questioned later on for political reasons.

5.5.4 Partial or Total Switch to Fully Funded Defined Contribution Schemes

Faced with the announced drift in the costs of Defined Benefit schemes and the political difficulty of reforming them "from inside", another strategy is for all or part of the first pillar to fully accept the character of Defined Contribution and abandon the Defined Benefit philosophy. For example, it is possible to retain only a first tier of Defined Benefit that provides a safety net. The benefits are either flat or calculated on a very low capped salary level. The remaining part of the contributions is transferred to a Defined Contribution system.

The idea of Defined Contribution fits quite naturally with a simultaneous shift from a Pay-As-You-Go scheme to a fully funded one. The principle is therefore to open individual savings accounts with mandatory deposits into which contributions are paid and invested on the financial markets. The savings accumulated at the time of retirement can then be converted into a life annuity providing a balanced pension benefit by definition. Many countries have set up such mandatory funded schemes (Sweden, Australia, Chile, Singapore, etc.). The modalities may vary mainly in terms of public or private management of the funds and in terms of freedom of investment choice for members. Some schemes also provide a guarantee since the switch to Defined Contribution places the risk of the financial markets on the beneficiaries (in the form of a minimum guaranteed rate of return or a minimum rate relative to the market).

This mechanism of "first pillar bis" in Defined Contribution and full funding has many advantages:

- it can coexist with a more modest basic Defined Benefit plan. Each country can decide the level of the cursor between these two levels according to its history and context,
- it is built on a principle of actuarial fairness and is not threatened by major demographic imbalance (only the risk of longevity after retirement needs special attention),
- it avoids imbalances caused by early retirement policies. On the contrary, it encourages work at advanced ages,
- it makes the individual responsible for the building of his or her retirement. Contributions are no longer taxes levied by a Welfare State but a form of long-term investment. This aspect is further reinforced when the affiliate can arbitrate the mode of investment himself,
- it is a stable, long-term form of saving in the national economy.

Regardless of the past ideological overtones against the very idea of full funding that is still present in some minds, the main difficulty in implementation may be the transition cost inherent in a partial transition from a Pay-As-You-Go scheme to a fully funded scheme. This argument undoubtedly has its value in the case of a complete transition to a fully funded scheme (as in Chile). On the other hand, a partial transition to a Defined Contribution fully funded scheme, if accompanied by various lighter parametric measures, can perfectly avoid significant transition costs, as the Swedish example shows.

5.5.5 *Notional Accounts and Point Systems*

Another approach developed in several countries to avoid this double funding syndrome in the event of a shift to a fully funded scheme is to try to combine Pay-As-You-Go philosophy and Defined Contribution.

Schemes such as point systems or notional accounts are based on this principle: financing is always on a Pay-As-You-Go basis, benefits are calculated not using an explicit salary formula, but correspond to a fictitious accumulated value of contributions paid throughout the career. This is therefore only a virtual calculation since there are no provisions to be invested on the financial markets. Countries such as Sweden, Latvia, Poland and Italy have chosen this mechanism for their first pillar.

The main advantage of the system is that it distills actuarial fairness into a Pay-As-You-Go plan. Everyone is expected to get back in benefits what they have contributed. Benefits are thus automatically correctly adjusted in the event of early retirement. But it contains the potential for imbalance inherent in a Pay-As-You-Go system.

Some hybrid schemes between DB and DC can also be considered. Different actuarial techniques of this kind will be presented and analyzed in Chaps. 6–8.

5.6 Buffer Funds and Leveling Techniques

As part of the management of their public pension schemes, several countries have set up a balanced fund to pre-fund the expected increase in the contribution rate in the coming years. These funds are generally fed using various cost-leveling techniques. This section explains the actuarial principles underlying these methods. It should be noted that these leveling techniques will also be used in the third part devoted to fully funding.

5.6.1 General Principles

The strict application of the Pay-As-You-Go method can lead, in the context of a Defined Benefit plan, to significant variations in expenses from one year to the next. It is therefore natural to try to smooth the effects: this is precisely the aim of the level and equalization methods. These techniques, which are applied on a financing method in order to equalize its effects, can be applied in a very general way to any method. Here we illustrate their application to a Pay-As-You-Go scheme in the context of the management of social security schemes.

The leveling principle is based on two tools:

- a smoothing process, substituting a stable cost in absolute or relative terms to an irregular cost (most often, stability is expressed in terms of the rate of contribution),
- a fund acting as a "clearing house".

The technique is particularly applicable when projections of future costs show a very strong expected growth in future contributions. The demographic changes to which the social security systems will be subject in many countries in the coming decades could justify the use of such a technique.

5.6.2 The Leveling Method

We consider a Defined Benefit plan to be funded on a pure Pay-As-You-Go basis. The future evolution of the plan's income and expenditure is projected at $t = 0$, based on demographic and financial assumptions, for a fixed period of time N. We use the following notation:

5.6 Buffer Funds and Leveling Techniques

- $SS(s)$ $(s = 0, 1, \ldots, N-1)$ the total salary at time s,
- $SR(s)$ $(s = 0, 1, \ldots, N-1)$ the total benefits to pay at time s,
- $\pi_{PA}(s)$ = the Pay-As-You-Go contribution rate given by Eq. (4.6):

$$\pi_{PA}(s) = \frac{SR(s)}{SS(s)}.$$

We want to substitute the variable contribution vector for a constant one, i.e. turn

$$(\pi_{PA}(0), \pi_{PA}(1), \ldots, \pi_{PA}(N-1)) \quad \text{into} \quad (\bar{\pi}, \bar{\pi}, \ldots, \bar{\pi}).$$

The principle of actuarial equivalence between the two contribution vectors is then applied. We obtain (denoting by $v = 1/(1+i)$ the discount factor):

$$\bar{\pi} \cdot \sum_{t=0}^{N-1} SS(t) \cdot v^t = \sum_{t=0}^{N-1} \pi_{PA}(t) \cdot SS(t) \cdot v^t$$

or

$$\bar{\pi} = \frac{\sum_{t=0}^{N-1} \pi_{PA}(t) SS(t) v^t}{\sum_{t=0}^{N-1} SS(t) v^t}. \tag{5.1}$$

The level rate is therefore the weighted average of the rates by the discounted salaries. Denoting by $C_N(s)$ the contribution to be made at time s for the level method with horizon N, we have

$$C_N(s) = \bar{\pi} \cdot SS(s).$$

The difference between this contribution and the contribution of a Pay-As-You-Go scheme feeds a *leveling fund* whose evolution is given by:

$$F(0) = (\bar{\pi} - \pi_{PA}(0)) \cdot SS(0)$$
$$F(1) = F(0) \cdot (1+i) + (\bar{\pi} - \pi_{PA}(1)) \cdot SS(1)$$
$$F(k) = F(k-1) \cdot (1+i) + (\bar{\pi} - \pi_{PA}(k)) \cdot SS(k) \quad (1 \leq k \leq N-1).$$

At the end of the level period ($k = N$), this level fund is equal to zero:

$$F(N) = (\bar{\pi} - \pi_{PA}(N-1)) \cdot SS(N-1) \cdot (1+i) + \cdots$$
$$+ (\bar{\pi} - \pi_{PA}(0)) SS(0) \cdot (1+i)^N$$
$$= \bar{\pi} \cdot \left(\sum_{j=0}^{N-1} SS(j)(1+i)^{N-j} \right) - \sum_{j=0}^{N-1} \pi_{PA}(j) SS(j)(1+i)^{N-j}$$
$$= (1+i)^N \cdot \left[\bar{\pi} \cdot \left(\sum_{j=0}^{N-1} SS(j)(1+i)^{-j} \right) - \sum_{j=0}^{N-1} \pi_{PA}(j) SS(j)(1+i)^{-j} \right]$$
$$= 0$$

by definition of $\bar{\pi}$.

When the contribution vector $(\pi_{PA}(0), \pi_{PA}(1), \ldots, \pi_{PA}(N-1))$ is arbitrary, the level fund can become negative, and the method cannot be applied. Therefore, this checking calculation should be performed before applying the method.

The method is particularly intended for the case of contribution rates increasing over time:

$$\pi_{PA}(0) < \pi_{PA}(1) < \cdots < \pi_{PA}(N-1).$$

This is obviously the standard case where the method finds all its *raison d'être*: the application of a level rate $\bar{\pi}$ (with $\pi_{PA}(0) < \bar{\pi} < \pi_{PA}(N-1)$) leads to paying more in the first years to avoid having to pay too much in the following years.

Let us consider, as an example, the simple case where salaries and pension benefits are increasing geometrically, with benefits increasing faster than salaries:

$$SR(t) = SR(0) \cdot (1+k)^t$$
$$SS(t) = SS(0) \cdot (1+r)^t$$

with $r < k$, so that

$$\pi_{PA}(t) = \frac{SR(t)}{SS(t)} = \frac{SR(0)}{SS(0)} \left(\frac{1+k}{1+r}\right)^t = \pi_0 (1+g)^t,$$

where

$$\pi_0 = \frac{SR(0)}{SS(0)} \quad \text{and} \quad 1+g = \frac{1+k}{1+r} > 1.$$

The contribution rates are therefore increasing geometrically. The level rate is then given by

$$\bar{\pi} = \frac{\sum_{t=0}^{N-1} \pi_0 (1+g)^t \cdot SS(0)(1+r)^t \cdot \left(\frac{1}{1+i}\right)^t}{\sum_{t=0}^{N-1} SS(0)(1+r)^t \left(\frac{1}{1+i}\right)^t}$$

$$= \pi_0 \cdot \frac{\left(1-a^N\right)/(1-a)}{\left(1-b^N\right)/(1-b)},$$

where $a = \frac{1+k}{1+i}$ and $b = \frac{1+r}{1+i}$.

5.6 Buffer Funds and Leveling Techniques

Table 5.2 Evolution of the balanced fund

Year	Original contribution rate in PAYG	Leveled contribution rate	Amount in the balanced fund
0	20.0%	21.74%	1.74
1	20.4%	21.74%	3.22
2	20.8%	21.74%	4.42
3	21.2%	21.74%	5.27
4	21.6%	21.74%	5.74
5	22.0%	21.74%	5.78
6	22.4%	21.74%	5.34
7	22.9%	21.74%	4.24
8	23.3%	21.74%	2.51
9	23.8%	21.74%	0
10	24.2%		

Example 5.1 Let us take $\pi_0 = 0,20$, $SS_0 = 100$, $k = 5\%$, $r = 3\%$, and $i = 6\%$. We're looking at the evolution of the scheme over the next 20 years, leveled over 10 years.

1. Level rate over the period [0; 10]:

$$\bar{\pi}_0 = 0.20 \cdot \frac{\left(1 - a^{10}\right)/(1 - a)}{\left(1 - b^{10}\right)/(1 - b)} = 0.2174.$$

2. The evolution of the contributions and the fund over [0; 10] is given in Table 5.2. At the beginning of year 10, the leveling fund is nil (end of the first leveling period), and the Pay-As-You-Go contribution rate is equal to 24.2%.

3. Leveled rate over the period [10; 20]:

$$\bar{\pi}_{10} = 0.242 \cdot \frac{\left(1 - a^{10}\right)/(1 - a)}{\left(1 - b^{10}\right)/(1 - b)} = 0.2631.$$

The leveling method we have applied therefore leads, at the end of the leveling reference period, to a very significant increase in the rate, from 21.74 to 26.31%, whereas its primary objective was to stabilize charges. Working with a fixed period of 10 years is obviously responsible for this phenomenon.

5.6.3 The Equalization Method

In order to avoid the problem of discontinuity at the end of the level period, the equalization method consists, each year when calculating the leveled rate, of gradually shifting the projection horizon in order to maintain a view over a fixed period of N years in the future. The level process is thus made dynamic.

In the first year, the method is identical to the leveling method (cf. Eq. (5.1)):

$$\bar{\pi}_0 = \frac{\sum_{t=0}^{N-1} \pi_{\text{PA}}(t) SS(t) v^t}{\sum_{t=0}^{N-1} SS(t) v^t}.$$

The fund at the end of the first year is given by:

$$F\left(1^-\right) = (\bar{\pi}_0 - \pi_{\text{PA}}(0)) \cdot SS(0) \cdot (1+i).$$

The actuarial equivalence principle is then applied over the period $[1, N+1]$:

Discounted value of leveled contributions + Fund

$$=$$

Discounted value of contributions in Pay-As-You-Go

i.e.

$$\bar{\pi}_1 \cdot \sum_{t=1}^{N} SS(t) v^{t-1} + F\left(1^-\right) = \sum_{t=1}^{N} SS(t) \cdot \pi_{\text{PA}}(t) \cdot v^{t-1}$$

or

$$\bar{\pi}_1 = \frac{\sum_{i=1}^{N} \pi_{\text{PA}}(t) SS(t) v^{t-1} - F\left(1^-\right)}{\sum_{t=1}^{N} SS(t) v^{t-1}}.$$

Generally speaking, we have at the kth year:

$$\bar{\pi}_k = \frac{\sum_{t=k}^{N+k-1} \pi_{\text{PA}}(t) SS(t) v^{t-k} - F\left(k^-\right)}{\sum_{t=k}^{N+k-1} SS(t) v^{t-k}},$$

with

$$F\left(k^-\right) = F\left((k-1)^-\right) + ((\bar{\pi}_{k-1} - \pi_{\text{PA}}(k-1)) SS(k-1))(1+i).$$

5.6 Buffer Funds and Leveling Techniques

Example 5.2 Going back to the previous assumptions, we have:

1. First year:

$$\bar{\pi}_0 = 0.2174$$

i.e. the same rate as with the leveling technique,

2. Second year:

$$F\left(1^{-}\right) = 1.74 \times 1.06 = 1.844$$

$$\bar{\pi}_1 = \frac{20 \sum_{t=1}^{10} \frac{(1+k)^t}{(1+i)^{t-1}} - 1.844}{100 \sum_{t=1}^{10} \frac{(1+r)^t}{(1+i)^{t-1}}} = 0.2196.$$

In addition to the advantage of gradually shifting the projection horizon—thus avoiding an abrupt discontinuity at the end of the leveling period—the method also makes it possible during the projection to capture any changes in relation to the assumptions made initially, such as

- an effective rate of return of the fund different from i,
- a future change in the i rate,
- a new estimate of future salary or benefit vectors.

The recalculation of the rate can be done not annually, but regularly, every T years. For instance we may consider a level method over ten years with recalculation every three years.

Chapter 6
Notional Defined Contribution Schemes

Abstract Notional Defined Contribution Schemes (NDCs) have been one of the major recent innovations for public social security schemes. The aim of this chapter is to present the logic behind NDCs and to study their main actuarial properties.

6.1 Basic Actuarial Mechanism

The notional accounts mechanism (also called a Notional Defined Contribution or NDC scheme) is based on a dual logic: financing is on a Pay-As-You-Go basis, and the calculation of benefits follows a logic of Defined Contribution. Everything happens as if we were working with individual funding, even if this calculation is virtual since the financing remains on a Pay-As-You-Go basis ("we act as if").

The calculation of the pension can be broken down into three steps: the notional capital, the conversion of this notional capital into an annuity, and the revaluation of the annuity.

6.1.1 Notional Capital

The notional capital corresponds to the notional accumulated value of all the contributions paid to the scheme during the member's career, these contributions being calculated at a constant contribution rate (by the DC nature of the system). We will assume that the contributions are paid at the beginning of the period, and we will use the following notation:

- $S_r(x)$ = total real salary received by the member at age x,
- $S(x)$ = the member's salary taken into account for his retirement,
- x_0 = first contribution age,
- x_r = legal retirement age,
- π = contribution rate,

- r_x = notional rate of return applied between age x and age $x + 1$ (virtual rate of return),
- $NC(z)$ = notional capital accumulated at age z (virtual provision at age z).

The contribution at age x paid to the member's notional capital will then be given by $\pi S(x)$.

It should be noted that this contribution to be included in the capital does not necessarily refer to the total salary of the scheme member. It is possible to introduce a social security cap, denoted P, for example, the part of the salary exceeding this cap being no longer taken into account in the contribution:

$$S(x) = \min(S_r(x), P).$$

We can also introduce a minimum salary, denoted $S_{\min}$, to take into account:

$$S(x) = \max(\min(S_r(x), P), S_{\min}).$$

The contributions generate a notional capital at any age corresponding to their sum capitalized at successive notional rates. Table 6.1 illustrates the progression of the notional account throughout the career. Generally speaking, the notional capital obtained at age z is given by

$$NC(z) = \sum_{x=x_0}^{z-1} \left(\pi S(x) \prod_{y=x}^{z-1} (1 + r_y) \right).$$

In particular, at legal retirement age, the notional capital is equal to

$$NC(x_r) = \sum_{x=x_0}^{x_r-1} \left(\pi S(x) \prod_{y=x}^{x_r-1} (1 + r_y) \right).$$

Regarding the parameters and this calculation, the following observations can be made:

Table 6.1 Progressive accumulation of the notional account

Age	Contribution	Notional rate	Notional capital (year-end)
x_0	$\pi S(x_0)$	r_{x_0}	$\pi S(x_0)(1 + r_{x_0})$
$x_0 + 1$	$\pi S(x_0 + 1)$	r_{x_0+1}	$\pi S(x_0)(1 + r_{x_0})(1 + r_{x_0+1})$ $+ \pi S(x_0 + 1)(1 + r_{x_0+1})$
...	...	...	...
z	$\pi S(z)$		$\sum_{x=x_0}^{z} \left(\pi S(x) \prod_{y=x}^{z} (1 + r_y) \right)$
...	...	...	...
$x_r - 1$	$\pi S(x_r - 1)$	r_{x_r-1}	$\sum_{x=x_0}^{x_r-1} \left(\pi S(x) \prod_{y=x}^{x_r-1} (1 + r_y) \right)$

1. As regards salaries, it should be noted that the system is designed to take into account all the salaries of the career (and not only the salaries of the last few years as in many DB schemes). As already observed, the salary taken into account in the formula does not always necessarily correspond to the total salary received by the scheme member. Indeed, one can consider minimum and maximum salaries in the formula; one can also include assimilated periods where a fictitious salary is taken into account (periods of unemployment or illness). The contribution actually paid by the member does not necessarily correspond to the contribution taken into account in the calculation of his notional capital. It should be noted that the virtual nature of the technique and the use of Pay-As-You-Go make it easier to provide solidarity than a with fully funded scheme.
2. Regarding the calculation of notional capital, we note that it is expressed as a simple accumulation of capitalized contributions, without taking into account mortality throughout the career. Most notional account schemes work in this way; we will see in Sect. 6.8 how mortality can be integrated before retirement. Let us remark that mortality could be included in the notional rate r (as the notation of the notional rate suggests, since it can depend on age!).
3. Regarding the virtual capitalization rate r, it should be noted that this rate, called the notional rate and variable from one year to the next, is a parameter to be fixed by the system. In traditional funding schemes, it is sufficient to use the observed financial rates of return on assets; in notional accounts, there are no assets and it is therefore necessary to choose a reference other than financial market rates. As we shall see, these rates will be in general linked to salary growth rates. On the other hand, even if theoretically these rates can depend explicitly on the age of the member, they are generally the same in any given year for all members. Sections 6.5 and 6.6 will discuss the choice of these notional rates in more detail.

6.1.2 Conversion of the Notional Capital into an Annuity

At retirement age, as in a traditional funding plan, the capital is converted into a lifetime annuity. The first retirement annuity at retirement age x_r is given by

$$R(x_r) = NC(x_r) G, \tag{6.1}$$

where $R(x_r)$ is the first retirement annuity at age, x_r, $NC(x_r)$ is the notional capital value at retirement and G is a conversion factor.

This conversion factor G is calculated by mimicking the principles of funding; it is based on the price of an indexed lifetime annuity.

Let us introduce the following notation:

- $p(x_r, x)$ = probability of survival at age x if still alive at retirement age x_r,
- ω = ultimate age of survival,
- j = annual pension revaluation rate (projected rate),
- r = notional discount rate (projected rate).

Assuming annuities payable annually in advance, the conversion factor is given by the following relationship:

$$\frac{1}{G} = \ddot{a}_{x_r} = \sum_{x=x_r}^{\omega} p(x_r, x) \frac{(1+j)^{x-x_r}}{(1+r)^{x-x_r}}. \tag{6.2}$$

This conversion rate depends on the following parameters:

1. The estimated survival probabilities after retirement, which may change from one generation of retirees to the next, depending on the estimated longevity adjustments specific to each cohort.
2. A rate of revaluation of pensions after retirement: we shall see that several choices are possible at this level; it is easy to see that a higher level of revaluation will lead to a smaller conversion factor G and therefore a lower initial pension. A natural choice is to index pensions to average salaries and thus to take the expected growth rate of individual salaries as the g factor. This canonical choice will be developed in Sect. 6.5.
3. A notional discount rate, reflecting the expected return of the system. The determination of this coefficient will also be detailed in Sect. 6.5.

The conversion rate can also be written:

$$\frac{1}{G} = \sum_{x=x_r}^{\omega} \frac{p(x_r, x)}{(1+a)^{x-x_r}},$$

with $1 + a = \frac{1+r}{1+j}$ and a = net discount rate.

In particular, when this net discount rate a is equal to zero (equality between the notional discount rate and the revaluation rate), the conversion coefficient corresponds to the inverse of the life expectancy at retirement age:

$$\frac{1}{G} = \sum_{x=x_r}^{\omega} p(x_r, x) = e_{x_r}.$$

Equation (6.1), which defines the pension benefit in notional accounts, can be put explicitly in the following form:

$$\sum_{x=x_0}^{x_r-1} \left(\pi S(x) \prod_{y=x}^{x_r-1} (1+r_y) \right) = R(x_r) \sum_{x=x_r}^{\omega} p(x_r, x) \frac{(1+j)^{x-x_r}}{(1+r)^{x-x_r}}. \tag{6.3}$$

This expresses the equality of the expected discounted values of the contribution flows and the benefit flows for a member throughout the life cycle.

6.1.3 Revaluation of the Retirement Annuity

After retirement, the pensions will be revalued each year according to an index which will be established in accordance with the choice of the revaluation coefficient j chosen in the calculation of the conversion rate G (see (6.2)). Thus, in general, we will have

$$R(x+1) = R(x)(1+j_x) \qquad (x > x_r),$$

where $R(x)$ is the retirement pension at age x and j_x is the real rate of revaluation between age x and age $x+1$.

For example, if we choose, in the conversion factor G, an expected revaluation coefficient j corresponding to the expected growth rate of the average salary, we will revalue pensions each year at the observed growth rate of the average salary.

6.2 Actuarial Neutrality

A natural property of the notional accounts technique is its actuarial neutrality in the event of an early or late retirement age (which is rarely the case with traditional Defined Benefit social security schemes). Indeed, Formula (6.3) for calculating the benefit written at legal retirement age x_r can be applied similarly at another effective retirement age.

If the member retires at another age, denoted x^* (with $x^* < x_r$ in case of early retirement and $x^* > x_r$ in case of postponement), the principle of equality of the current values of contributions and benefits is expressed at that age and Formula (6.3) becomes:

$$\sum_{x=x_0}^{x^*-1} \left(\pi S(x) \prod_{y=x}^{x^*-1} (1+r_y) \right) = R\left(x^*\right) \sum_{x=x^*}^{\omega} p\left(x^*, x\right) \frac{(1+j)^{x-x^*}}{(1+r)^{x-x^*}}. \qquad (6.4)$$

Denoting by $NC(x^*)$ the notional capital accumulated at the effective age x^* of retirement, we thus have

$$R\left(x^*\right) = \frac{1}{\ddot{a}_{x^*}} NC\left(x^*\right).$$

Compared to what happens with the normal retirement age, a correction coefficient is therefore applied:

$$R\left(x^*\right) = G\beta_{x^*} NC\left(x^*\right), \qquad (6.5)$$

with $\beta_{x^*} = \frac{\ddot{a}_{x_r}}{\ddot{a}_{x^*}}$.

In the case of pension at legal age, the retirement pension would have been equal to

$$R(x_r) = G\, NC(x_r). \tag{6.6}$$

In particular:

- in the case of early retirement ($x^* < x_r$), $\beta_{x^*} < 1$: the benefit is reduced (on an actuarially sound basis),
- in the case of postponement ($x^* > x_r$), $\beta_{x^*} > 1$: the benefit is increased (direct incentive to postpone retiring).

Note that the comparison of the pensions obtained at the normal retirement age and at any age x^*, respectively formulas (6.6) and (6.5), highlights a double correction. On the one hand we have a correction coefficient that allows us to take into account the fact that the pension will be paid for a longer or shorter period of time according to the pension's starting date. On the other hand, the notional capital NC will be different depending on the age at which the contributions end. For example, in the case of early retirement, there will be fewer contributions than expected and they will be capitalized at the notional rates for a shorter period of time, these two phenomena leading to a notional capital lower than the one which would have been obtained in the case of continued activity until the legal retirement age.

These effects illustrate the actuarial neutrality of notional accounts. From this perspective, let us explicitly calculate in notional accounts the effect of bringing forward retirement by one year.

In case of retirement at age x_r, the notional retirement pension is given explicitly by

$$R(x_r) = \frac{\sum_{x=x_0}^{x_r-1}\left(\pi S(x) \prod_{y=x}^{x_r-1}(1+r_y)\right)}{\ddot{a}_{x_r}}.$$

Similarly, in the case of departure one year earlier at age $x_r - 1$, we have

$$R(x_r - 1) = \frac{\sum_{x=x_0}^{x_r-2}\left(\pi . S(x) \prod_{y=x}^{x_r-2}(1+r_y)\right)}{\ddot{a}_{x_r-1}}.$$

By comparing these two successive values, we obtain

$$R(x_r - 1) = \frac{\sum_{x=x_0}^{x_r-2}\left(\pi S(x) \left(\prod_{y=x}^{x_r-1}(1+r_y)\right)\right)}{\ddot{a}_{x_r}} \frac{\ddot{a}_{x_r}}{\ddot{a}_{x_r-1}} \frac{1}{(1+r_{x-1})}$$

$$= \frac{\ddot{a}_{x_r}}{\ddot{a}_{x_r-1}} \frac{1}{(1+r_{x_r-1})} (R(x_r) - M),$$

6.3 Notional Accounts: DC or DB?

with

$$M = \frac{1}{\ddot{a}_{x_r}} \pi S(x_r - 1)(1 + r_{x_r - 1}).$$

A one-year anticipation of retirement affects the amount of the benefit at three levels:

$$R(x_r - 1) = R_2 R_3 (R(x_r) - M),$$

where

- $R_2 = \frac{\ddot{a}_{x_r}}{\ddot{a}_{x_r - 1}}$ corrects for the effect that the pension will be paid out one more year; as with funding, this correction is expressed as a ratio of two annuities but this time calculated in the notional bases (cf. (6.2))
- The ratio $R_3 = \frac{1}{(1 + r_{x_r - 1})}$ corrects for the fact that the pension will be paid one year earlier; as in funding, this is a discount factor over one year, using here only the notional rate for the year. In this term, there is no mortality effect because the presented calculation of notional account did not take into account pre-retirement mortality. As already mentioned above, we will show in Sect. 6.8 how the pre-retirement survival effect can be integrated into notional accounts.
- $M = \frac{1}{\ddot{a}_{x_r}} \pi S(x_r - 1)(1 + r_{x_r - 1})$ corrects for the effect of one less contribution.

This actuarial neutrality is much more problematic in traditional Defined Benefit social security schemes:

- While the penalty is implicit in a notional scheme, it must be made explicit in a Defined Benefit scheme and may therefore be perceived very negatively by the members.
- The penalty is often linear (reduction in proportion to the number of years of membership compared to the theoretical number of years until retirement age) and not actuarially sound.
- There are few incentives to postpone retirement in traditional Defined Benefit systems.

Finally, it should be noted that the concept of legal retirement age loses some of its importance in a notional accounts system. Indeed, the general equivalence formula (6.4) can be applied at any age. These systems therefore offer a great deal of flexibility, combined with a sense of responsibility from the member through the way his or her retirement benefits are calculated.

6.3 Notional Accounts: DC or DB?

While the notional accounts technique looks like a Defined Contribution plan, it also has similarities with a Defined Benefit type system that takes into account the salaries of the entire career (average career indexed system).

In fact, in a DB scheme based on a revalued career average (see Sect. 2.3.1.1), the pension benefit of an individual i at retirement age can be written

$$R(x_r) = \frac{\alpha}{(x_r - x_0)} \sum_{x=x_0}^{x_r-1} S(x)h(x), \tag{6.7}$$

where $h(x)$ is the salary revaluation function between age x and retirement, and α is the percentage awarded by the plan.

On the other hand, the generic formula (6.1) for notional accounts leads to the pension benefit

$$R(x_r) = \frac{1}{\ddot{a}_{x_r}} \pi \sum_{x=x_0}^{x_r-1} S(x) \left(\prod_{j=x}^{x_r-1} (1+r_j) \right). \tag{6.8}$$

The equivalence between expressions (6.7) and (6.8) is done naturally by choosing, for the salary revaluation function

$$h(x) = \prod_{j=x}^{x_r-1} (1+r_j)$$

and for the percentage granted by the plan

$$\alpha = \frac{\pi (x_r - x_0)}{\ddot{a}_{x_r}} = \pi G (x_r - x_0).$$

Therefore, a notional account system can be identified with a system of Defined Benefit based on average revalued salaries, where the revaluation function is based on virtual capitalization rates, and the benefit rate is the product of the contribution rate, the conversion coefficient and the full length of the career. In particular, in notional account systems where the virtual funding rate is linked to salary growth, we obtain a classic revaluation function based on a salary growth index.

While the notional accounts system may therefore appear to be another way of formulating a career average Defined Benefit scheme, we should nevertheless note the adaptive nature of the notional technique: whereas in a traditional Defined Benefit scheme, the percentage α granted by the plan is an exogenous and fixed parameter of the system, the equivalence formula (6.3) shows that in notional accounts, this coefficient will take into account, through the conversion coefficient G, the evolution of life expectancy.

On the other hand, the notional accounts technique also has, by definition and contrary to a Defined Benefit scheme, an objective of contribution rate stability. The actuarial equilibrium conditions and automatic adjustment mechanisms presented

6.4 Notional Accounts and Replacement Rate

in the following will illustrate this tension between DB and DC inherent in notional accounts.

Finally, it should be noted that in the context of increasing flexibility in retirement age, we have shown in Sect. 6.2 that notional accounts imply actuarial neutrality in contrast to Defined Benefit systems.

6.4 Notional Accounts and Replacement Rate

The replacement rate is an important instrument for measuring the quality of a pension scheme. It consists of a ratio comparing the first pension paid on retirement with the salaries earned prior to retirement. It is possible to compare the pension to the last active salary, but it is also possible to compare the pension to an average of salaries over several years, or even over the entire career.

If we use the most commonly used reference, based on the last working salary, we can express the replacement rate at any age x^* by the relation

$$rr(x^*) = \frac{R(x^*)}{S(x^*)} = \frac{1}{\ddot{a}_{x^*}} \pi \sum_{x=x_0}^{x^*-1} \left(\frac{S(x)}{S(x^*)} \prod_{j=x}^{x^*-1} (1+r_j) \right).$$

Taking into account the explicit form of the conversion factor:

$$rr(x^*) = \frac{\pi \sum_{x=x_0}^{x^*-1} \left(\frac{S(x)}{S(x^*)} \prod_{j=x}^{x^*-1} (1+r_j) \right)}{\sum_{x=x^*}^{\omega} p(x^*, x) \frac{(1+j)^{x-x^*}}{(1+r)^{x-x^*}}}.$$

In particular, if we assume that the salaries increase each year by a constant growth rate, denoted g, we obtain

$$rr(x^*) = \frac{\pi \sum_{x=x_0}^{x^*-1} \left((1+g)^{x-x^*} \prod_{j=x}^{x^*-1} (1+r_j) \right)}{\sum_{x=x^*}^{\omega} p(x^*, x) \frac{(1+j)^{x-x^*}}{(1+r)^{x-x^*}}}.$$

This replacement rate can therefore vary over time and will differ from one individual to another. It depends on the one hand on parameters specific to the individual (length of service, effective retirement age x^*, salary growth rate) and on the other hand on parameters of the scheme (notional rate r, revaluation rate g, contribution rate, survival probability).

Table 6.2 illustrates the link between the replacement rate and these different parameters. It explains and comments on the effect on the replacement rate when, all other things being equal, the parameter being studied increases.

Table 6.2 Dependence of the replacement rate on various parameters in notional accounts

Variable (symbol)	Effect of growth on replacement rate	Comment
Effective retirement age (x^*)	Increasing	Retiring later has a double effect (larger notional capital and shorter benefit period)
Starting age (x_0)	Decreasing	Starting younger allows for more contributions for a longer period
Salary growth rate (g)	Decreasing	A high growth rate devalues the oldest contributions paid in the past
Rate of pension revaluation (j)	Decreasing	Greater revaluation of pensions reduces the initial level
Notional rate (r)	Increasing	Funding at a higher rate increases the notional account and decreases the discounted value of pensions
Probability of survival ($p^*(x^*, x)$)	Decreasing	Better survival (longevity risk) increases the discounted value of pensions

6.5 Canonical Choice of Notional Accounts for a Steady-State Scheme

Expression (6.3), which defines the level of pensions in notional accounts at the legal retirement age, is based on a logic of equivalence at the level of the individual, following the example of what is applied in a funded Defined Contribution scheme. Indeed, in the latter case, the retirement pension for the individual will be the solution of the equation

$$\sum_{x=x_0}^{x_r-1} \left(\pi S(x) \prod_{y=x}^{x_r-1} \left(1 + i_y\right) \right) = R(x_r) \sum_{x=x_r}^{\omega} p(x_r, x) \frac{(1+j)^{x-x_r}}{(1+i)^{x-x_r}}, \qquad (6.9)$$

where i_y is the effective financial rate of return at age y and i is the financial discount rate.

Relationships (6.3) and (6.9) seem perfectly similar; yet there is an important difference between these two systems:

- In the fully funding methods, Formula (6.9) not only defines the pension to be paid, but also expresses the actuarial equivalence resulting from the use of individual funding (effective provision at retirement age); the methods are therefore automatically actuarially balanced since, by definition of fully funding, equilibrium is observed by comparing contributions and benefits over a career;
- In notional accounts, the actuarial equilibrium of the method is based on Pay-As-You-Go, i.e. on a collective equilibrium between contribution income

6.5 Canonical Choice of Notional Accounts for a Steady-State Scheme

and pension expenditure observed in the same year for the entire population. However, the definition of the notional account pension, while it imitates the logic of funding by adapting its parameters, does not seem to take this collective actuarial balance into account at all.

There is therefore no guarantee that a notional accounts system will be balanced, and in general it will not be, except in special cases corresponding in particular to clever choices of parameters (and in particular notional rates).

In this section, we will place ourselves in a stationary regime where the parameters of the model remain constant over time. Under this assumption, we will develop the so-called canonical choice of parameters which leads to a natural actuarial equilibrium. Section 6.6 will generalize these results to the case of a non-stationary demographic and financial environment.

Let us first define the stationary framework considered here. We work with a uniform average salary by age and a uniform average pension. Each year, salaries and pensions are indexed at a fixed rate g. Let $S(x, t)$ denote the average salary at time t for the age cohort x. The evolution of the salary for a given member from one year to the next is influenced on the one hand by a salary progression linked to age and on the other hand by a revaluation linked to general salary revaluation. We will denote by $\varphi(x)$ the salary progression associated with age x. The salary of an individual of age x at time t, denoted $S(x, t)$, will thus be given by

$$S(x, t) = S_{x_0}\varphi(x)(1 + g)^t.$$

Entry into the population is at the uniform entry age x_0. Retirement is at uniform age x_r. The effect of mortality before retirement is ignored. After retirement, the survival probabilities are stationary, so let $p(x_r, x)$ be the survival probability at age x if still alive at retirement age x_r. Each year, the population size grows at any age by a factor d.

At time t, the number of new entrants into the population at age x_0 will be denoted

$$L(x_0, t) = L_0(1 + d)^t. \tag{6.10}$$

The cohort with age x at time t (x being pre-retirement) entered the population at time $t - (x - x_0)$; its size will be given by

$$L(x, t) = L_0(1 + d)^{t-(x-x_0)} \quad (x_0 \leq x \leq x_r). \tag{6.11}$$

Similarly, for ages after retirement age, we will have

$$L(x, t) = L_0(1 + d)^{t-(x-x_0)} p(x_r, x) \quad (x > x_r). \tag{6.12}$$

Let us develop a notional account system in this population. We first consider the generation retiring at age x_r at time t. For this generation, the different career salaries can be obtained as a function of the final salary by correcting for the effects of inflation and salary progressions:

$$S(x) = S(x, t + (x - x_r)) = S_{x_0}\varphi(x)(1+g)^{t+(x-x_r)}.$$

Let $S_f(t)$ be the salary at time t at the last working age $x_r - 1$ and revalued at time t:

$$S_f(t) = S(x_r - 1, t - 1)(1+g) = S_{x_0}\varphi(x_r - 1)(1+g)^t.$$

We can write

$$S(x, t + (x - x_r)) = S_f(t) \frac{\varphi(x)}{\varphi(x_r - 1)} \frac{1}{(1+g)^{x_r - x}}$$

$$= S_f(t)\Psi(x)\frac{1}{(1+g)^{x_r - x}},$$

defining

$$\Psi(x) = \frac{\varphi(x)}{\varphi(x_r - 1)} = \frac{S(x, t)}{S(x_r - 1, t)} = \frac{S(x, t)}{S_f(t)},$$

the ratio between the salary of an individual of age x and the salary at the same time of an individual just before retirement (correction for the salary progression).

Formula (6.8) of the pension in notional accounts thus gives for this generation a first pension equal to

$$R(x_r, t) = \frac{\pi \sum_{x=x_0}^{x_r-1} S(x)(1+r)^{x_r-x}}{\ddot{a}_{x_r}} = \frac{\pi \sum_{x=x_0}^{x_r-1} S_f(t) \frac{(1+r)^{x_r-x}}{(1+g)^{x_r-x}} \Psi(x)}{\sum_{x=x_r}^{\omega} p(x_r, x) \frac{(1+g)^{x-x_r}}{(1+r)^{x-x_r}}}.$$

(6.13)

Similarly, the previous generation, which retired one year ago at time $t - 1$, had a first pension, given by

$$R(x_r, t-1) = \frac{\pi \sum_{x=x_0}^{x_r-1} S_f(t-1) \frac{(1+r)^{x_r-x}}{(1+g)^{x_r-x}} \Psi(x)}{\sum_{x=x_r}^{\omega} p(x_r, x) \frac{(1+g)^{x-x_r}}{(1+r)^{x-x_r}}}.$$

At time t, this annuity is indexed by a factor $(1+g)$; it thus becomes

$$R(x_r+1, t) = R(x_r, t-1)(1+g) = \frac{\pi \sum_{x=x_0}^{x_r-1} S_f(t-1) \frac{(1+r)^{x_r-x}}{(1+g)^{x_r-x}} \Psi(x)}{\sum_{x=x_r}^{\omega} p(x_r, x) \frac{(1+g)^{x-x_r}}{(1+r)^{x-x_r}}} (1+g).$$

6.5 Canonical Choice of Notional Accounts for a Steady-State Scheme

Given that the salaries also evolve at the rate g, we have

$$S_f(t-1)(1+g) = S_f(t)$$

and therefore

$$R(x_r+1, t) = \frac{\pi \sum_{x=x_0}^{x_r-1} S_f(t) \frac{(1+r)^{x_r-x}}{(1+g)^{x_r-x}} \Psi(x)}{\sum_{x=x_r}^{\omega} p(x_r, x) \frac{(1+g)^{x-x_r}}{(1+r)^{x-x_r}}} = R(x_r, t).$$

In general, given the stationarity conditions of the model and the assumption that pensions are indexed to salaries, all generations will receive at time t the same retirement pension given by

$$R(z, t) = \frac{\pi \sum_{x=x_0}^{x_r-1} S_f(t) \frac{(1+r)^{x_r-x}}{(1+g)^{x_r-x}} \Psi(x)}{\sum_{x=x_r}^{\omega} p(x_r, x) \frac{(1+g)^{x-x_r}}{(1+r)^{x-x_r}}} = R(t) \quad \forall z \geq x_r. \tag{6.14}$$

Note that this formula does not seem to depend on the rate of population growth d, which may seem surprising for Pay-As-You-Go funding.

In fact, the Pay-As-You-Go technique has not yet been used to obtain these pension benefits. We will now express the actuarial equilibrium constraint, which states that in pure Pay-As-You-Go, the sum of the benefits paid at time t to all retirees must correspond to the sum of the contributions taken from the workers at that same time. This equilibrium relation will allow us to highlight an equilibrated notional rate.

Let us therefore calculate at time t, on the one hand, the aggregate of contributions, and on the other hand, the aggregate of benefits. The aggregate of contributions is equal to the average contribution multiplied by the number of active members at time t:

$$\pi \sum_{x=x_0}^{x_r-1} L(x, t) S(x, t) = \pi \sum_{x=x_0}^{x_r-1} L(x, t) S_f(t) \Psi(x).$$

The benefit aggregate is the average pension multiplied by the number of retirees at time t:

$$R(t) \sum_{x=x_r}^{\omega} L(x, t).$$

Equating the two aggregates and taking into account Formula (6.14) of the retirement pension $R(t)$ and the population relationships (6.11) and (6.12), we obtain

$$\pi S_f(t) \sum_{x=x_0}^{x_r-1} L_0(1+d)^{t-(x-x_0)} \Psi(x)$$

$$= \left(\frac{\pi S_f(t) \sum_{x=x_0}^{x_r-1} \left(\frac{1+r}{1+g}\right)^{x_r-x} \Psi(x)}{\sum_{x=x_r}^{\omega} p(x_r,x) \left(\frac{1+g}{1+r}\right)^{x-x_r}} \right) \sum_{x=x_r}^{\omega} L_0(1+d)^{t-(x-x_0)} p(x_r,x)$$

or, after regrouping and simplification,

$$\frac{\sum_{x=x_0}^{x_r-1}(1+d)^{t-(x-x_0)} \Psi(x)}{\sum_{x=x_r}^{\omega} p(x_r,x)(1+d)^{t-(x-x_0)}} = \frac{\sum_{x=x_0}^{x_r-1} \left(\frac{1+r}{1+g}\right)^{x_r-x} \Psi(x)}{\sum_{x=x_r}^{\omega} p(x_r,x) \left(\frac{1+g}{1+r}\right)^{x-x_r}}$$

or, by multiplying the top and bottom of the left-hand expression by $(1+d)^{x_r-t-x_0}$,

$$\frac{\sum_{x=x_0}^{x_r-1}(1+d)^{x_r-x} \Psi(x)}{\sum_{x=x_r}^{\omega} p(x_r,x)(1+d)^{x_r-x}} = \frac{\sum_{x=x_0}^{x_r-1} \left(\frac{1+r}{1+g}\right)^{x_r-x} \Psi'(x)}{\sum_{x=x_r}^{\omega} p(x_r,x) \left(\frac{1+r}{1+g}\right)^{x_r-x}}.$$

Actuarial equilibrium will therefore be achieved if the following condition is met:

$$1+r = (1+d)(1+g). \tag{6.15}$$

This fundamental formula shows us that, under the conditions of the model, it is appropriate to choose as notional rate r the rate given by Formula (6.15); this rate represents the composition of the growth rate of the population and the growth rate of the average individual salary. This rate therefore corresponds to the rate of growth of the mass of contributions (accumulation of a demographic effect and a salary effect).

This is the way the canonical model of notional accounts is generated, leading to the following double rule: current pensions are indexed at the same rate as the average individual salary, and notional accounts are capitalized at the growth rate of the mass of contributions.

In this canonical model, the retirement annuity becomes:

$$R(t) = \frac{\pi \sum_{x=x_0}^{x_r-1} S_f(t) \frac{1}{(1+d)^{x-x_r}} \Psi(x)}{\sum_{x=x_r}^{\omega} p(x_r,x) \frac{1}{(1+d)^{x-x_r}}}. \tag{6.16}$$

Two remarks must be made. On the one hand, this canonical choice leads to equilibrium of the notional regime only under the assumptions of stationarity of the model. In particular, it assumes that mortality does not change over time and follows the law initially predicted, that the population remains in the stationary state described by relations (6.11) and (6.12), and that the salary, which is uniform for all, evolves each year at a constant rate. We will see in Sect. 6.7 that, when these hypotheses are no longer fulfilled, the use of the canonical choice, even if it remains intuitively natural, no longer guarantees actuarial equilibrium in general. Other rules will have to be applied or automatic adjustment mechanisms will have to be put in place.

On the other hand, we have assumed that there is no pre-retirement mortality (virtual pre-retirement savings account mechanism); this assumption can be relaxed and the effects of mid-career deaths on the mechanics of the plan can be taken into account. This point will be addressed in Sect. 6.8.

6.6 Variants of the Canonical Choice for a Steady-State Scheme

In the stationary model, the canonical choice is a sufficient condition of equilibrium but it is not a necessary condition. Other schemes can also lead to actuarial equilibrium. The main variants concern the way in which the pension is indexed after retirement. In the canonical model presented above, we have assumed that the pension grows each year at the same rate as the average salary of the working population. The conversion rate used in the denominator of the formula uses revaluation projected at the rate of growth of the average salary. This approach seems natural insofar as it allows retirees' incomes to evolve in the same way as those of active workers. But one could also revaluate at another rate, or not revaluate at all, provided that one remains consistent between the way in which the first retirement pension is calculated (conversion factor) and the way in which the pension is revaluated in practice afterwards.

For example, suppose we decide to index pensions at a rate j different from the growth rate g of average salary. In this case, the formulas developed above must be adapted at two levels:

- Value of the first pension for the new retirees: Formula (6.13) must be modified to account for another notional capital conversion factor and now becomes

$$R(x_r, t) = \frac{\pi \sum_{x=x_0}^{x_r-1} S(x)(1+r)^{x_r-x}}{\ddot{a}_{x_r}} = \frac{\pi \sum_{x=x_0}^{x_r-1} S_f(t) \frac{(1+r)^{x_r-x}}{(1+g)^{x_r-x}} \Psi(x)}{\sum_{x=x_r}^{\omega} p(x_r, x) \frac{(1+j)^{x-x_r}}{(1+r)^{x-x_r}}}.$$

- Revaluation of past pensions: the pensions of previous generations must be indexed at the rate j and not g. Thus, for the generation retired for one year at time t, Formula (6.14) becomes

$$R(x_r+1,t) = R(x_r,t-1)(1+j)$$

$$= \frac{\pi \sum_{x=x_0}^{x_r-1} S_f(t-1)\frac{(1+r)^{x_r-x}}{(1+g)^{x_r-x}}\Psi(x)}{\sum_{x=x_r}^{\omega} p(x_r,x)\frac{(1+g)^{x_r-x}}{(1+r)^{x-x_r}}}(1+j)$$

$$= R(x_r,t)\left(\frac{1+j}{1+g}\right).$$

In general, each generation of retirees will have a different pension, and Formula (6.14) becomes for a retiree of age x_r+m in year t:

$$R(x_r+m,t) = \frac{\pi \sum_{x=x_0}^{x_r-1} S_f(t-m)\frac{(1+r)^{x_r-x}}{(1+g)^{x_r-x}}\Psi(x)}{\sum_{x=x_r}^{\omega} p(x_r,x)\frac{(1+j)^{x-x_r}}{(1+r)^{x-x_r}}}(1+j)^m$$

$$= \frac{\pi \sum_{x=x_0}^{x_r-1} \frac{S_f(t)}{(1+g)^m}\frac{(1+r)^{x_r-x}}{(1+g)^{x_r-x}}\Psi(x)}{\sum_{x=x_r}^{\omega} p(x_r,x)\frac{(1+j)^{x-x_r}}{(1+r)^{x-x_r}}}(1+j)^m$$

$$= R(x_r,t)\left(\frac{1+j}{1+g}\right)^m.$$

The aggregate of pensions to be paid in year t is then equal to

$$\sum_{x=x_r}^{\omega} L(x,t)R(x,t).$$

The balance between incomes and expenses in Pay-As-You-Go is now written

$$\pi \sum_{x=x_0}^{x_r-1} L(x,t)S_f(t)\Psi(x) = \sum_{x=x_r}^{\omega} L(x,t)R(x,t),$$

i.e.

$$\pi S_f(t) \sum_{x=x_0}^{x_r-1} L_0(1+d)^{t-(x-x_0)}\Psi(x)$$

$$= \sum_{x=x_r}^{\omega} L_0(1+d)^{t-(x-x_0)} p(x_r,x) \frac{\pi \sum_{y=x_0}^{x_r-1} \frac{(1+r)^{x_r-y}}{(1+g)^{x_r-y}}\Psi(x)}{\sum_{y=x_r}^{\omega} p(x_r,y)\frac{(1+j)^{y-x_r}}{(1+r)^{y-x_r}}}\left(\frac{1+j}{1+g}\right)^{x-x_r},$$

6.6 Variants of the Canonical Choice for a Steady-State Scheme

which can be written

$$\frac{\sum_{x=x_0}^{x_r-1}(1+d)^{x_r-x}\Psi(x)}{\sum_{x=x_r}^{\omega}p(x_r,x)(1+d)^{x-x_r}\left(\frac{1+j}{1+g}\right)^{x-x_r}}$$

$$=\frac{\sum_{x=x_0}^{x_r-1}\left(\frac{1+r}{1+g}\right)^{x_r-x}\Psi(x)}{\sum_{x=x_r}^{\omega}p(x_r,x)\left(\frac{1+g}{1+r}\right)^{x-x_r}\left(\frac{1+j}{1+g}\right)^{x-x_r}}.$$

The actuarial equilibrium condition on the notional rate r is therefore the same as (6.15):

$$1+r=(1+d)(1+g).$$

The choice of the revaluation parameter j for post-retirement pensions is an important parameter of flexibility in a notional account system. In particular, a value of the parameter j that is lower than the growth rate g of salaries leads, through a more favorable conversion factor, to an initial retirement pension that is higher than in the canonical choice, but that grows less quickly thereafter. There is thus a trade-off between the initial level and the growth rate. In the stationary model considered here, this choice does not influence the equilibrium value of the notional rate r, which remains equal to the growth rate of the mass of contributions.

The canonical choice leads to a pension revaluation rate of $j=g$. In this case, the conversion annuity of the notional capital takes the form

$$\ddot{a}_{x_r}=\sum_{x=x_r}^{\omega}p(x_r,x)\left(\frac{1+j}{1+r}\right)^{x-x_r}=\sum_{x=x_r}^{\omega}p(x_r,x)\frac{1}{(1+d)^{x-x_r}}. \qquad (6.17)$$

The conversion annuity is calculated as a constant lifetime annuity under a discount rate corresponding to the population growth rate.

Alternatively, one could consider the following cases:

1. $j=0$: no revaluation of the pensions. In this case, the conversion annuity becomes

$$\ddot{a}_{x_r}=\sum_{x=x_r}^{\omega}p(x_r,x)\frac{1}{(1+r)^{x-x_r}}$$

(constant lifetime annuity discounted at the notional rate).
2. $j=g-1\%$: revaluation at the rate of salary growth minus a fixed percentage (e.g., indexing pensions 1% less than salaries). Cases 1 and 2 lead to a higher initial pension than in the canonical case.

3. $1 + j = 1 + r = (1 + g)(1 + d)$: revaluation of pensions at the notional rate (rate of growth of the mass of contributions). In this case, the revaluation will be more favorable (respectively more unfavorable) than in the canonical model when the population is growing (respectively decreasing). The annuity then takes the simple form of a life expectancy at retirement:

$$\ddot{a}_{x_r} = \sum_{x=x_r}^{\omega} p(x_r, x). \qquad (6.18)$$

4. $1 + j = \frac{1+r}{1.01}$: revaluation of pensions at the notional rate corrected by a spread of 1%. In this case, the annuity becomes

$$\ddot{a}_{x_r} = \sum_{x=x_r}^{\omega} p(x_r, x) \frac{1}{(1,01)^{x-x_r}}$$

(constant lifetime annuity discounted at the spread rate chosen here at 1%).

6.7 A Three-Period Non-Stationary Equilibrium Model

The actuarial equilibrium model considered in the previous section was based on demographic and financial stationarity assumptions. The purpose of this section is to examine whether an equilibrium is still possible in the presence of a non-stationary model. To illustrate the point, we will work with a simple three-period model that allows us to stylize the two important phenomena in notional accounts: the capitalization of the account at the notional rate during working life and the revaluation of pensions after retirement. We will see in particular how to adapt in this non-stationary model the choices of parameters to be made in order to maintain an actuarial balance in PAYG.

6.7.1 Assumptions

At the demographic level, members enter the population at age y, work one period; they are retired at age $y + 1$ and receive a first pension at that age. At age $y + 2$, survivors receive a second pension. At age $y + 3$, there are no more survivors.

At time t, the number of new entrants into the population at age y will be denoted

$$L(y, t) = L_0 \prod_{j=1}^{t} (1 + d_j). \qquad (6.19)$$

6.7 A Three-Period Non-Stationary Equilibrium Model

The population growth rate, d_j, is assumed to vary from year to year (compared to the stationary assumption (6.10)).

At retirement age $y + 1$, everyone is assumed to have survived and we have

$$L(y+1, t) = L(y, t-1) = L_0 \prod_{j=1}^{t-1} (1 + d_j). \tag{6.20}$$

Similarly at the next age $y + 2$, we have

$$L(y+2, t) = L(y+1, t-1) p_t = L_0 p_t \prod_{j=1}^{t-2} (1 + d_j). \tag{6.21}$$

The probability of surviving between age $y + 1$ and age $y + 2$ at time t, denoted p_t, can evolve over time (for example as an increasing function of time to model increasing longevity).

From age $y + 3$, there are no more survivors:

$$L(z, t) = 0 \quad \forall z \geq y + 3.$$

In this population, workers of age y receive a salary given at time t by

$$S(t) = S_0 \prod_{j=1}^{t} (1 + g_j).$$

The salary growth rate g_t and the pension growth rate j_t are also assumed to be variable from one year to the next, unlike in the stationary model.

6.7.2 Notional Accounts Scheme

Let us develop a notional system in this population based on a constant contribution rate π. For the generation retiring at time t at age $y + 1$, the pension to be paid will be of the form

$$R(y+1, t) = \frac{\pi S(t-1)(1 + r_t)}{a(t)}.$$

The numerator of this fraction represents the notional account at retirement (fictitious capitalization of the contributions); the denominator is the conversion annuity to apply at time t. We have to choose two parameters: the notional rate of the year t, denoted r_t, and the conversion annuity to be applied, denoted $a(t)$.

For individuals of age $y+2$ at time t, the pension obtained when they retire at $t-1$, can similarly be written

$$R(y+1, t-1) = \frac{\pi S(t-2)(1+r_{t-1})}{a(t-1)}.$$

At time t this pension is revalued and becomes

$$R(y+2, t) = R(y+1, t-1)(1+j_t) = \frac{\pi S(t-2)(1+r_{t-1})}{a(t-1)}(1+j_t).$$

Let us calculate at time t the aggregate of contributions and benefits:

- aggregate of contributions:

$$\pi S(t) L(y, t),$$

- aggregate of benefits:

$$R(y+1, t) L(y+1, t) + R(y+2, t) L(y+2, t).$$

By equating these two expressions, we obtain

$$\pi S(t) L_0 \prod_{j=1}^{t}(1+d_j) = \frac{\pi S(t-2)(1+r_{t-1})}{a(t-1)}(1+j_t) L_0 p_t \prod_{j=1}^{t-2}(1+d_j)$$

$$+ \frac{\pi S(t-1)(1+r_t)}{a(t)} L_0 \prod_{j=1}^{t-1}(1+d_j). \qquad (6.22)$$

After simplification, we get

$$(1+g_{t-1})(1+g_t)(1+d_{t-1})(1+d_t)$$
$$= \frac{1+r_{t-1}}{a(t-1)}(1+j_t) p_t + \frac{(1+r_t)(1+g_{t-1})}{a(t)}(1+d_{t-1}). \qquad (6.23)$$

This can still be written by expressing the pension revaluation rate as a function of the other parameters:

$$1+j_t$$
$$= \frac{(1+g_{t-1})(1+d_{t-1}) a(t-1)}{p_t} \frac{1+r_t}{a(t)} \frac{1+r_t}{1+r_{t-1}} \left(\frac{(1+g_t)(1+d_t)}{(1+r_t)} a(t) - 1 \right).$$
$$(6.24)$$

6.7 A Three-Period Non-Stationary Equilibrium Model

This equilibrium relationship includes three parameters to be calibrated in year t: the notional rate r_t, the pension revaluation rate j_t and the conversion annuity $a(t)$, based on the observed evolution of the three variables g (observed salary growth), d (observed population growth) and p (observed survival probability). There are thus an infinite number of possible systems.

6.7.3 Stationary Special Case

Let us first observe that in the stationary case (all parameters becoming independent of time), Relationship (6.24) reduces to

$$1 + j = \frac{(1+g)(1+d)}{p} \left(\frac{(1+g)(1+d)}{(1+r)} a - 1 \right).$$

Taking into account the value of the notional rate (6.15) of equilibrium:

$$1 + j = \frac{(1+r)}{p}(a - 1)$$

or

$$a = 1 + p\frac{1+j}{1+r},$$

which corresponds to the notional conversion annuity in the model considered here.

This confirms that in a stationary regime, the use of the natural notional rate (6.15) allows us to obtain equilibrium, whatever the value of the revaluation rate of pensions. In particular, pensions can be indexed to the average salary (canonical choice).

6.7.4 Non-Stationary General Case

Let us show that, in general, the canonical choice is no longer actuarially balanced. In this dynamic model, the canonical choice would correspond to the following parameterization:

- Notional rate: observed growth in total contributions:

$$1 + r_t = (1 + d_t)(1 + g_t), \tag{6.25}$$

- Pension revaluation rate: observed individual salary growth:

$$1 + j_t = 1 + g_t, \qquad (6.26)$$

- Conversion annuity: annuity price based on observable data at time t:

$$a(t) = 1 + p_t \frac{1 + g_t}{1 + r_t} = 1 + p_t \frac{1}{1 + d_t}. \qquad (6.27)$$

Let us then replace in Expression (6.24) the notional rate with the proposed value (6.25); we get

$$1 + j_t = (1 + r_t) \frac{a(t-1)}{a(t)} \frac{1}{p_t} (a(t) - 1). \qquad (6.28)$$

Taking into account the value of the annuity (6.27), we obtain

$$1 + j_t = (1 + g_t) \frac{a(t-1)}{a(t)} = (1 + g_t) \kappa_t. \qquad (6.29)$$

In this non-stationary case, therefore, a corrective coefficient appears in relation to the canonical choice (6.26): given the notional rate and the form of the annuity, the revaluation of pensions no longer exactly follows that of salaries; an adjustment coefficient appears, which is expressed in the form of the ratio between two annuities:

$$\kappa_t = \frac{a(t-1)}{a(t)}. \qquad (6.30)$$

In particular, if we assume a stable population growth rate in year t ($d_{t-1} = d_t = d$), we obtain:

$$\kappa_t = \frac{1 + \frac{p_{t-1}}{1+d}}{1 + \frac{p_t}{1+d}}. \qquad (6.31)$$

If longevity improves ($p_{t-1} < p_t$), the adjustment coefficient is less than 1. Pensions can now only be indexed at a lower rate than salaries; there is an automatic adjustment linked to the increase in longevity.

An increase in the working population would be necessary in parallel to neutralize this longevity effect. Indeed, given (6.29), the condition $a(t-1) = a(t)$, allowing for the identical revaluation of pensions and salaries, would give

$$1 + d_t = (1 + d_{t-1}) \frac{p_t}{p_{t-1}}.$$

6.7 A Three-Period Non-Stationary Equilibrium Model

In conclusion, and contrary to the stationary case, the canonical choice (6.25), (6.26), (6.27), although still intuitive, no longer ensures actuarial equilibrium in a non-stationary model. Thus, in the model considered, if we wish to maintain the logic of the notional rate (6.25) and conversion annuity (6.27) formulas, the pension revaluation formula (6.26) cannot be maintained but must become (6.29).

6.7.5 Variants

The solution (6.29) developed above is a sufficient condition for equilibrium; it is not a necessary condition. We will develop here three variants based on the following alternative adaptations:

- Variant 1: use of prospective probabilities,
- Variant 2: revaluation based on the notional rate,
- Variant 3: adjustment of the notional rate.

6.7.5.1 Variant 1: Use of Prospective Probabilities

The equilibrium parameters obtained, derived from the canonical choice (notional rate (6.25), annuity (6.27), and pension revaluation rate (6.29)), depend only on elements that are observable at time t; this makes their determination easy, since they are based on measurable quantities. It should be noted in particular that the annuity is based on the probability observed at time t (periodic table) and not on a projected probability at $t+1$ (table of the cohort considered), as is sometimes used in certain notional account models.

The use of observed probabilities is actually more natural in Pay-As-You-Go since the actuarial equilibrium is considered at a single instant, and not over the whole life of a cohort. In a funding context, it would have been more natural to use prospective probabilities per cohort. This rule can obviously also be considered here, given the number of degrees of freedom. But we will show that this parameterization, although possible, leads to a more complex and counter-intuitive behavior of the indexing mechanism.

For example, if we use the following values for the first two parameters:

- notional rate:

$$1 + r_t = (1 + d_t)(1 + g_t),$$

- conversion annuity:

$$a(t) = 1 + p_{t+1} \frac{1}{1 + d_t}.$$

(conversion annuity using a forward-looking probability rather than an observed probability), the equilibrium Eq. (6.28) gives

$$1 + j_t = (1 + g_t) \frac{a(t-1)}{a(t)} \frac{p_{t+1}}{p_t}. \quad (6.32)$$

Compared to the adjustment coefficient (6.30), there appears therefore a complementary term based on the ratio of two successive probabilities:

$$\kappa_t = \frac{a(t-1)}{a(t)} \frac{p_{t+1}}{p_t}.$$

In addition to its more complex form than Formula (6.30), let us show that this rule, applied in a context of increasing longevity, leads to a counter-intuitive correction. Let us assume, as before, that population growth is stable ($d_{t-1} = d_t = d$); in this case, Relationship (6.32) can be written

$$\kappa_t = \frac{p_{t+1}}{p_t} \frac{1 + \frac{p_t}{1+d}}{1 + \frac{p_{t+1}}{1+d}} \quad (6.33)$$

In the case of improved longevity (i.e. $p_t < p_{t+1}$), it is easy to show that now the adjustment coefficient is greater than one:

$$\frac{p_{t+1}}{p_t} \frac{1 + \frac{p_t}{1+d}}{1 + \frac{p_{t+1}}{1+d}} > 1 \Leftrightarrow p_{t+1}\left(1 + \frac{p_t}{1+d}\right) > p_t\left(1 + \frac{p_{t+1}}{1+d}\right)$$

$$\Leftrightarrow p_{t+1} > p_t.$$

This would lead to increasing pensions at a higher rate than the rate of growth of salaries when longevity increases in the population! The rule obtained by (6.31) thus seems more logical; the use of current probabilities rather than prospective probabilities is thus to be recommended. One can also interpret these results as follows: the adjustment coefficients (6.31) and (6.33) suggest that in the case of increasing longevity:

- Using current probabilities leads to an initial pension that is too generous and must be less revaluated than salaries in order to safeguard the actuarial balance,
- Using prospective probabilities leads to an initial pension that is too low and must be more revaluated than salaries.

6.7.5.2 Variant 2: Revaluation Based on the Notional Rate

If the revaluation of pensions is done, not on the basis of the growth rate of salaries, but on the basis of the growth rate of the salary mass, the conversion annuity

6.7 A Three-Period Non-Stationary Equilibrium Model

becomes a life expectancy (cf. (6.27)), and using current probabilities, we have

$$a(t) = 1 + p_t.$$

The notional rate is always given by

$$1 + r_t = (1 + d_t)(1 + g_t).$$

The equilibrium Eq. (6.28) then becomes

$$1 + j_t = (1 + r_t) \frac{a(t-1)}{a(t)} \frac{1}{p_t} (a(t) - 1).$$

Substituting the annuity by its value, we get

$$1 + j_t = (1 + r_t) \kappa_t.$$

The actual revaluation is therefore based on the desired revaluation (here the notional rate) but corrected again by the same annuity ratio as in the adaptation (6.29):

$$\kappa_t = \frac{a(t-1)}{a(t)}.$$

6.7.5.3 Variant 3: Simultaneous Adjustment of the Notional Rate and the Revaluation Rate

So far, we have only applied a correction factor to the pension revaluation rate. One could also correct the notional rate. We will assume that the notional rate can be written using the canonical form (growth rate of the salary mass), corrected by an adjustment coefficient to be determined:

$$1 + r_t = (1 + g_t)(1 + d_t) \varphi_t. \tag{6.34}$$

We will assume that the same correction applies to the revaluation of current pensions:

$$1 + j_t = (1 + g_t) \varphi_t. \tag{6.35}$$

The conversion annuity is always given by

$$a(t) = 1 + p_t \frac{1}{1 + d_t}.$$

Substituting these values (6.34) and (6.35) into the general equilibrium relation (6.24), we get

$$1 = \frac{1}{p_t} \frac{a(t-1)}{a(t)} \frac{(1+d_t)}{\varphi_{t-1}} \left(\frac{a(t)}{\varphi_t} - 1 \right).$$

We then obtain a recurrence relationship for the correction factor on the standard notional rate:

$$\varphi_t = \frac{a(t)}{1 + p_t \frac{a(t)}{a(t-1)} \frac{\varphi_{t-1}}{1+d_t}} = \frac{1 + p_t \frac{1}{1+d_t}}{1 + p_t \frac{a(t)}{a(t-1)} \frac{\varphi_{t-1}}{1+d_t}}.$$

In particular, in the first year when the system ceases to be stationary (i.e. $\varphi_{t-1} = 1$), the first correction to be made takes the form:

$$\varphi_t = \frac{1 + p_t \frac{1}{1+d_t}}{1 + p_t \frac{a(t)}{a(t-1)} \frac{1}{1+d_t}}. \tag{6.36}$$

In particular, if the population growth rate d is stable, an increase in the survival probability p will lead to an increase in the conversion annuity:

$$\frac{a(t)}{a(t-1)} > 1$$

and thus by Relationship (6.36), to a first correction factor $\varphi < 1$.

6.7.5.4 Conclusion

In the non-stationary case (i.e. when the demographic growth rate, longevity and/or salary growth are not constant), the application of the classic canonical rule does not lead to actuarial equilibrium. We then have three variables on which it is possible to act in order to get back to this equilibrium: the pension revaluation rate, the notional rate and the conversion annuity. The results obtained in the model considered suggest the following recommendations:

- Several correction mechanisms are possible, given the three degrees of freedom available.
- The technique of notional accounts is more suited to the determination of conversion annuities using observed probabilities rather than prospective probabilities. This result may seem at first sight counter-intuitive to the idea of converting a notional capital for the retiring generation. But it is in fact perfectly consistent with the underlying financing mechanism, which remains purely Pay-As-You-Go and not generational funding.

- The correction factors obtained are generally expressed in terms of the ratio of two successive conversion annuities (cf. factors of the type (6.30)).

6.8 Integration of Mortality Profits

We have so far ignored the effect of pre-retirement mortality (post-retirement mortality being well taken into account in the conversion annuity of the notional account). The notional capital formula is just based on the capitalization of contributions at successive notional rates; the population models considered did not include pre-retirement mortality. This assumption is often used in notional models by analogy with the evolution of a standard savings account. But there is of course nothing to prevent the inclusion of pre-retirement mortality.

If we consider in this case that for contributors who die during their career, no acquired rights are retained, we can intuitively expect that considering mortality during the career will improve the performance of the notional system, following the example of the "mortality credit" phenomenon in life insurance. The contributions paid by members who die prematurely before retirement age do not give them any rights under this hypothesis; they will make it possible to improve the retirement of the survivors.

6.8.1 Introduction of the Survival Dividend

In order to understand the mortality credit and survival dividend phenomena on a simple example, let us first take up the 3 period model presented in Sect. 6.7, but revised this time in a stationary version.

In such a model, the size of the population changes according to the following pattern (see (6.19), (6.20) and (6.21)):

- labor force (age y):

$$L(y, t) = L_0(1 + d)^t,$$

- population at retirement age (age $y + 1$):

$$L(y + 1, t) = L(y, t - 1) = L_0(1 + d)^{t-1},$$

- population at last age (age $y + 2$):

$$L(y + 2, t) = L(y + 1, t - 1)p = L_0 p(1 + d)^{t-2},$$

where d is the population growth rate and p is the probability of survival after retirement.

Let us now also introduce a pre-retirement mortality. Let p_0 denote the probability of surviving during the career (between entry age y and retirement age $y+1$). We then have

$$L(y,t) = L_0(1+d)^t$$

$$L(y+1,t) = L_0(1+d)^{t-1} p_0$$

$$L(y+2,t) = L_0(1+d)^{t-2} p_0\, p.$$

The salaries of working people are supposed to follow the law:

$$S(t) = S_0(1+g)^t.$$

The pension paid to the generation retiring at time t is given by the notional capital capitalized at the new notional rate r^*, divided by the annuity:

$$R^*(y+1,t) = \frac{\pi S(t-1)(1+r^*)}{a}.$$

The conversion annuity is then given according to the canonical choice (see (6.27)) by

$$a = 1 + p\frac{1}{1+d}. \tag{6.37}$$

In order to determine the new notional equilibrium rate r^* and assuming that pensions are revaluated as salaries, the equilibrium relation (6.22) allows us to obtain the notional rate to be applied in order to obtain actuarial equilibrium:

$$\pi S(t)L(y,t) = R^*(y+1,t)L(y+1,t) + R^*(y+2,t)L(y+2,t).$$

By substituting, Expression (6.23) becomes this time

$$(1+g)^2(1+d)^2 = \frac{1+r^*}{a}(1+g)pp_0 + \frac{(1+r^*)(1+g)}{a}(1+d)p_0$$

or, after simplification, and taking into account the form of the conversion annuity (6.37),

$$1 + r^* = \frac{(1+g)(1+d)}{p_0}. \tag{6.38}$$

6.8 Integration of Mortality Profits

The notional rate is thus increased compared to the canonical version by the factor $1/p_0$. Taking into account the mortality in the model allows a higher notional rate to be granted to survivors.

In particular, we can highlight the concept of survival dividend at retirement age, denoted DS, defined as the difference between the notional capital with mortality credit and the notional capital without this effect:

$$DS(t) = NC^*(y+1,t) - NC(y+1,t), \qquad (6.39)$$

where $NC^*(y+1,t) =$ notional capital at retirement with mortality effect and $NC(y+1,t) =$ notional capital at retirement without mortality effect.

This dividend is given here by

$$DS(t) = \pi S(t-1)\left(r^* - r\right) = \pi S(t)(1+d)\left(\frac{1}{p_0} - 1\right). \qquad (6.40)$$

6.8.2 Partial Survival Dividend

Rather than redistributing the value of accounts resulting from pre-retirement deaths completely to survivors, a mixed allocation of accounts in the event of pre-retirement death could be provided:

- A proportion $1 - \alpha$ of the accounts joins the survival dividend for the cohort survivors as detailed above.
- A proportion α of the accounts serves as survival benefits, for example, for the surviving spouse; it does not, therefore, serve the overall revaluation of the cohort survivors' accounts.

In this case, as in Sect. 6.8.1, we can still highlight a notional rate increased by the effect of the mortality credit; but this time, only part of the mortality credit is allocated to the revaluation of the cohort's accounts.

6.8.3 Survival Dividend and Longevity Funding

Assuming that no survivor-type benefits are paid out before retirement, the existence of the mortality credit generated by members who contributed before retirement but died before retirement age yields an extra amount per cohort reaching retirement that we have called a survival dividend and that can be used in different ways.

6.8.3.1 Increase in Benefits

As suggested above, the survival dividend can be used totally to increase the retirement pensions of the cohort. In this case, the initial annuity at retirement without mortality credit, given by

$$R(y+1,t) = \frac{\pi S(t-1)(1+r)}{a} = \frac{\pi S(t)(1+d)}{a},$$

becomes

$$R^*(y+1,t) = \frac{\pi S(t-1)(1+r^*)}{a} = \frac{\pi S(t)(1+d)}{p_0 \, a}.$$

6.8.3.2 Pre-Funding of Longevity

In the absence of survival benefits, the existence of mortality credits makes it possible to increase the first retirement pension from the $R(y+1,t)$ level to the $R^*(y+1,t)$ level. But one could also alternatively keep the survival dividend and use it to finance a possible increase in survival after retirement.

In this case, the initial level of the pension $R(y+1,t)$ is maintained (the members are not credited directly) and the survival dividend makes it possible, in the model considered here, to cope with a better longevity than expected at age $y+2$, i.e. a real probability of survival between age $y+1$ and age $y+2$, of a level p^* higher than the expected level p (with $p^* > p$).

Let us then calculate the maximum survival increment that would be financed by the survival dividend (maximum level of probability p^*). In the absence of a longevity effect, the commitment at retirement age y for the cohort is the notional capital NC given by

$$R(y+1,t)a = \pi S(t)(1+d).$$

In the presence of a longevity effect between age $y+1$ and age $y+2$, the liabilities become $R(y+1,t)a^*$, where a^* is the new annuity computed using the increased survival probability p^* (compare to Eq. (6.37)):

$$a^* = 1 + p^* \frac{1}{1+d}.$$

The difference in liabilities due to increased longevity can therefore be written

$$\Delta(t) = \pi S(t)(1+d)\left(\frac{a^*}{a} - 1\right) = \pi S(t)(1+d)\left(\frac{1 + \frac{p^*}{1+d}}{1 + \frac{p}{1+d}} - 1\right). \quad (6.41)$$

6.8 Integration of Mortality Profits

The dividend makes it possible to finance this liability increase when $\Delta(t) \leq DS(t)$, i.e. given Eqs. (6.40) and (6.41):

$$\left(\frac{1 + \frac{p^*}{1+d}}{1 + \frac{p}{1+d}} - 1 \right) \leq \frac{1}{p_0} - 1.$$

This delivers the upper bound on post-retirement survival:

$$p^* \leq \frac{p}{p_0} + (1+d)\left(\frac{1}{p_0} - 1\right). \qquad (6.42)$$

As numerical example, suppose that the probability before retirement is $p_0 = 0.98$, the expected probability after retirement is $p = 0.9$ and the population growth rate $d = 0.01$. Then we obtain by applying Formula (6.42) a maximum probability of survival after retirement of $p^* = 0.939$.

6.8.4 General Multi-Age Model

In the multi-age stationary model as developed in Sect. 6.6, we generalize now the effects of pre-retirement mortality on the actuarial equilibrium of the plan. We introduce in this model survival probabilities between any couple of ages (and no longer only after retirement age).

We will denote by $p(y, x)$ the probability of survival at age x if still alive at age y ($y \leq x$). These probabilities are assumed to be independent of time in a stationary model. The population functions (6.11) and (6.12) are now both given before and after retirement age by

$$L(x, t) = L_0 p(x_0, x)(1+d)^{t-(x-x_0)} \quad (x_0 \leq x \leq \omega). \qquad (6.43)$$

We now define a notional accounts model where the notional rate can be age-dependent in order to consider age-dependent annual survival probabilities. Taking the generic notional accounts formula (6.13) and introducing age-dependent notional rates, we obtain for the generation retiring in year t

$$R(x_r, t) = \frac{\pi \sum_{x=x_0}^{x_r-1} S(x) \prod_{y=x}^{x_r-1}(1+r_y)}{\ddot{a}_{x_r}} = \frac{\pi \sum_{x=x_0}^{x_r-1} \frac{S_f(t)\Psi(x)}{(1+g)^{x_r-x}} \prod_{y=x}^{x_r-1}(1+r_y)}{\ddot{a}_{x_r}},$$

where r_y is the notional rate to be applied between ages y and $y+1$, the conversion annuity a being given by (6.17).

If we revaluate pensions each year at the same rate as salaries, we have seen that all generations in this stationary model will receive at time t the same pension (cf. Relationship (6.14)), which now becomes

$$R(z,t) = \frac{\pi \sum_{x=x_0}^{x_r-1} \frac{S_f(t)\Psi(x)}{(1+g)^{x_r-x}} \prod_{y=x}^{x_r-1} (1+r_y)}{\ddot{a}_{x_r}} = R(t) \quad \forall z \geq x_r. \tag{6.44}$$

In the model without taking into account pre-retirement mortality, we have shown that the system is balanced on a Pay-As-You-Go basis by choosing for the notional rate (see (6.15)):

$$1 + r = (1+d)(1+g).$$

Inspired by Formula (6.38), it seems natural to propose this time as notional rate r_y:

$$1 + r_y = \frac{(1+g)(1+d)}{p(y, y+1)}.$$

The product of notional factors then takes the simple form:

$$\prod_{y=x}^{x_r-1} (1+r_y) = \frac{(1+g)^{x_r-x}(1+d)^{x_r-x}}{\prod_{y=x}^{x_r-1} p(y, y+1)} = \frac{(1+g)^{x_r-x}(1+d)^{x_r-x}}{p(x, x_r)}.$$

Using this notional rate form, the retirement pension (6.44) becomes

$$R(t) = \frac{\pi \sum_{x=x_0}^{x_r-1} \frac{S_f(t)\Psi(x)}{(1+g)^{x_r-x}} \prod_{y=x}^{x_r-1} (1+r_y)}{\ddot{a}_{x_r}} = \frac{\pi \sum_{x=x_0}^{x_r-1} \frac{S_f(t)\Psi(x)(1+d)^{x_r-x}}{p(x,x_r)}}{\sum_{x=x_r}^{\omega} p(x_r, x) \left(\frac{1}{1+d}\right)^{x-x_r}}.$$

Let us show that there is an actuarial equilibrium, i.e. equality at the level of the whole population between the contributions received and the retirement pensions to be paid:

$$\pi \sum_{x=x_0}^{x_r-1} L(x,t) S_f(t) \Psi(x) = R(t) \sum_{x=x_r}^{\omega} L(x,t).$$

6.8 Integration of Mortality Profits

Using the form (6.43) of the population function $L(x, t)$, the right-hand term of this equation becomes

$$R(t) \sum_{x=x_r}^{\omega} L(x, t) = \frac{\pi \sum_{x=x_0}^{x_r-1} \frac{S_f(t)\Psi(x)(1+d)^{x_r-x}}{p(x,x_r)}}{\sum_{x=x_r}^{\omega} p(x_r x) \left(\frac{1}{1+d}\right)^{x-x_r}} \sum_{x=x_r}^{\omega} L_0 p(x_0, x)(1+d)^{t-(x-x_0)}$$

$$= \left(\pi \sum_{x=x_0}^{x_r-1} \frac{S_f(t)\Psi(x)(1+d)^{x_r-x}}{p(x, x_r)}\right) L_0(1+d)^t \cdot$$

$$(1+d)^{-(x_r-x_0)} p(x_0, x_r)$$

$$= \pi \sum_{x=x_0}^{x_r-1} S_f(t)\Psi(x) p(x_0, x) L_0 (1+d)^{t-(x-x_0)}$$

$$= \pi \sum_{x=x_0}^{x_r-1} S_f(t)\Psi(x) L(x, t).$$

There is therefore an actuarial balance.

Following the example of (6.39), a survival dividend can be introduced, defined as the difference between the lifetime capitalized value and the financial capitalized value of contributions over the entire career:

$$DS(t) = NC^*(x_r, t) - NC(x_r, t)$$

$$= \pi S_f(t) \sum_{x=x_0}^{x_r-1} \Psi(x)(1+d)^{x_r-x} \left(\frac{1}{p(x, x_r)} - 1\right). \tag{6.45}$$

This survival dividend can be compared to the cost of financing longevity, resulting from an increase in life expectancy at retirement age and therefore in the conversion annuity.

Let us assume that due to the effect of increased longevity, the conversion annuity increases from a level

$$\ddot{a}_{x_r} = \sum_{x=x_r}^{\omega} p(x_r, x) \left(\frac{1+g}{1+r}\right)^{x-x_r} = \sum_{x=x_r}^{\omega} p(x_r, x) \frac{1}{(1+d)^{x-x_r}}$$

to the level

$$\ddot{a}_{x_r}^* = \sum_{x=x_r}^{\omega} p^*(x_r, x) \left(\frac{1+g}{1+r}\right)^{x-x_r} = \sum_{x=x_r}^{\omega} p^*(x_r, x) \frac{1}{(1+d)^{x-x_r}}.$$

The total cost then becomes, assuming that mortality benefits are not credited to notional rates but retained in the survival dividend (see (6.41)),

$$\Delta(t) = \left(\pi S_f(t) \sum_{x=x_0}^{x_r-1} \Psi(x)(1+d)^{x_r-x} \right) \left(\frac{\ddot{a}^*_{x_r}}{\ddot{a}_{x_r}} - 1 \right). \tag{6.46}$$

Comparing relationships (6.45) and (6.46) suggests that the survival dividend will be able to finance the increase in life expectancy if the new conversion annuity resulting from the new life table satisfies the following bound condition:

$$\ddot{a}^*_{x_r} \leq \ddot{a}_{x_r} \frac{\sum_{x=x_0}^{x_r-1} \frac{\Psi(x)(1+d)^{x_r-x}}{p(x,x_r)}}{\sum_{x=x_0}^{x_r-1} \Psi(x)(1+d)^{x_r-x}}.$$

In particular, in the absence of salary effects and population growth, we obtain

$$\ddot{a}^*_{x_r} \leq \ddot{a}_{x_r} \frac{\sum_{x=x_0}^{x_r-1} \frac{1}{p(x,x_r)}}{x_r - x_0}.$$

6.9 Implicit Debt and Pay-As-You-Go Actuarial Balance Sheet

6.9.1 *Short-Term or Long-Term Equilibrium*

The actuarial equilibrium equation in Pay-As-You-Go (income = expenses each year) is by definition based on a short-term approach, in contrast to funding, where one works with discounted values of future flows generally over long periods of time. This does not mean that a longer-term approach is not desirable and feasible in Pay-As-You-Go. A government cannot be satisfied with a short-term view of the balance of its social security system alone. Future projections are desirable in order to manage the system. Establishing the balance of a Pay-As-You-Go scheme over the long term is, however, more delicate than in the case of funding, and several definitions are possible. Indeed, different concepts exist, making it possible to highlight, on the one hand, the implicit debt involved in any Pay-As-You-Go system and, on the other hand, the balance of the scheme. The difficulty in establishing a balance sheet for a Pay-As-You-Go scheme lies in its a priori infinite time horizon. Any attempt to consider flows only over a finite time horizon can therefore lead to distortions; definitions will diverge in terms of the populations taken into account in the calculation and in the way in which the horizon is truncated to make it finite (in practice, projections are always established over finite time horizons).

6.9 Implicit Debt and Pay-As-You-Go Actuarial Balance Sheet

Generally speaking, a scheme's liability (or provision), viewed prospectively, will take the following form:

Provision = discounted value of benefits − discounted value of contributions

We can then theoretically highlight the following four fundamental levels (the first three being on a finite horizon, the fourth on an infinite horizon):

1. Retirees' liability: only the rights of current pensioners are considered at the valuation date. The provision is then written as

   Provision = discounted value of future benefits of current retirees.

   If i is the chosen discount rate, one has

   $$V(t) = \sum_{x=x_r}^{\omega} \sum_{s=0}^{\omega-x} R(x+s, t+s) L(x+s, t+s) \frac{1}{(1+i)^s}.$$

2. Closed-system debt: we consider only the population (active or retired) existing at the valuation date and we project future contributions and future benefits for these two groups:

   Provision = discounted value of future benefits of current retirees
   + discounted value of future benefits of current workers
   − discounted value of future contributions of current workers.

   If i is the chosen discount rate, $p(x, y)$ the probability of surviving to age y having been alive at age x, and $S(x+s, t+s)$ is the average salary at age $x+s$ at time $t+s$,

   $$V(t) = \sum_{x=x_r}^{\omega} \sum_{s=0}^{\omega-x} R(x+s, t+s) L(x+s, t+s) \frac{1}{(1+i)^s}$$

   $$+ \sum_{x=x_0}^{x_r-1} L(x,t) \sum_{s=x_r-x}^{\omega-x} R(x+s, t+s) p(x, x+s) \frac{1}{(1+i)^s}$$

   $$- \pi \sum_{x=x_0}^{x_r-1} L(x,t) \sum_{s=0}^{x_r-x-1} S(x+s, t+s) p(x, x+s) \frac{1}{(1+i)^s}.$$

3. Debt in an open system over a finite horizon: we consider all present and future generations and we consider the projection of future benefits and contributions, but we limit ourselves to a specific time horizon of length T:

$$V(t) = \sum_{s=0}^{T} \sum_{x=x_r}^{\omega} R(x, t+s) L(x, t+s) \frac{1}{(1+i)^s}$$

$$- \pi \sum_{s=0}^{T} \sum_{x=x_0}^{x_r-1} S(x, t+s) L(x, t+s) \frac{1}{(1+i)^s}.$$

4. Debt in an open system on an infinite horizon: we consider all present and future generations on an infinite horizon, with no time limit:

$$V(t) = \sum_{s=0}^{\infty} \sum_{x=x_r}^{\omega} R(x, t+s) L(x, t+s) \frac{1}{(1+i)^s}$$

$$- \pi \sum_{s=0}^{\infty} \sum_{x=x_0}^{x_r-1} S(x, t+s) L(x, t+s) \frac{1}{(1+i)^s}.$$

Methods 3 and 4 in an open system obviously require assumptions to be made about the renewal of the population; methods 1 and 2 are based on the population existing at the time of the calculation. It should be noted that in the case of funding, the calculation of provisions is generally based on the philosophy of method 2 (current values of future flows of benefits and contributions for the population existing at the valuation date). In Pay-As-You-Go, given the implicit assumption of solidarity between successive generations, method 4 with an infinite horizon seems the most appropriate. In practice, the horizon is truncated and method 3 is used (projection of the plan over a given future horizon).

In these methods 3 and 4, the system is said to be in equilibrium if the provision $V(t)$ is zero; if this provision is strictly positive, we speak of an implicit debt. Some countries, such as Sweden, also use method 2. to assess the equilibrium of the scheme in order to be able to establish an actuarial balance sheet, as in the case of funding. Sections 6.9.2 and 6.9.3 below explain the logic used in this context.

6.9.2 Balance Sheet of a Closed System Scheme

While the notion of an actuarial balance sheet is natural in a funding environment (comparison between the market value of financial assets and the discounted value of liabilities), it may seem more difficult to grasp in a Pay-As-You-Go context characterized precisely by an absence of financial funds (or with a residual "buffer"

6.9 Implicit Debt and Pay-As-You-Go Actuarial Balance Sheet

Table 6.3 Balance sheets for different types of pension schemes

(a)

Assets	Liabilities
Financial Assets	Provisions for retirees
	Provisions for contributors

(b)

Assets	Liabilities
Contribution asset	Provisions for retirees
	Provisions for contributors

(c)

Assets	Liabilities
Contribution asset	Provisions for retirees
Buffer fund	Provisions for contributors

fund). However, monitoring the solvency of a Pay-As-You-Go pension scheme in the same way as a funding system and being able to detect imbalances are just as necessary. One way of adapting a scheme's balance sheet to the case of Pay-As-You-Go when working in a closed system (see Method 2 in Sect. 6.9.1) is to add an item on the asset side that precisely offsets the liabilities when the plan is in equilibrium in the steady state (a balance sheet item sometimes called a "contribution asset").

In a funding scheme, the balance sheet of a pension scheme takes the classic form given in Table 6.3a. In a Pay-As-You-Go plan with no initial funds, the balance sheet of a pension fund becomes that given in Table 6.3b. When there is a buffer fund in addition in a Pay-As-You-Go system, the balance sheet becomes that given in Table 6.3c.

In order to obtain the value of the contribution asset, let us consider a steady-state notional account regime based on the canonical model (see Sect. 6.5) and compute its closed-system provisions. We therefore assume:

- a population whose evolution is given by
  - before retirement age:

$$L(x,t) = L_0(1+d)^{t-(x-x_0)} \quad (x_0 \leq x \leq x_r),$$

  - after retirement age:

$$L(x,t) = L_0(1+d)^{t-(x-x_0)} p(x_r, x) \quad (x > x_r),$$

- an average salary at age x and time t given by

$$S(x,t) = S_0(1+g)^t . \Psi(x) = S_f(t)\Psi(x)$$

(with the convention $\Psi(x_r - 1) = 1$),

- a notional rate r given by the relation

$$1 + r = (1+d)(1+g),$$

- a notional account pension given in the canonical model by (see (6.16)):

$$R(t) = \frac{\pi \sum_{x=x_0}^{x_r-1} S_f(t) \frac{1}{(1+d)^{x-x_r}} \Psi(x)}{\sum_{x=x_r}^{\omega} p(x_r, x) \frac{1}{(1+d)^{x-x_r}}}.$$

Let us calculate the discounted value of future liabilities. All these discounted values will be calculated here using a discount rate equal to the notional rate r used in the benefit computation.

The total value of the liabilities, which will be calculated prospectively, can be broken down into two parts: that relating to members already retired and that relating to members still contributing (cf. Relationship (4.7)):

$$V(t) = V_r(t) + V_a(t).$$

1. Present value of retirees' liabilities:

$$V_r(t) = \text{discounted value of future benefits}$$
$$\text{of members already retired at time } t$$

$$= R(t) \sum_{x=x_r}^{\omega} \sum_{s=0}^{\omega-x} L(x+s, t+s) \frac{(1+g)^s}{(1+r)^s}$$

$$= R(t) \sum_{x=x_r}^{\omega} L(x, t) \left(\sum_{s=0}^{\omega-x} p(x, x+s) \frac{1}{(1+d)^s} \right)$$

$$= R(t) \sum_{x=x_r}^{\omega} L(x, t) \ddot{a}_x.$$

Note that this global provision of the scheme can be expressed as a function of the sum of the individual provisions that correspond to the value of the notional account of each retiree (retrospective view):

$$V_r(t) = \sum_{x=x_r}^{\omega} L(x, t) (R(t) \ddot{a}_x) = \sum_{x=x_r}^{\omega} L(x, t) NC(x, t),$$

with $NC(x, t)$ = notional account of a retiree of age x at time t.

6.9 Implicit Debt and Pay-As-You-Go Actuarial Balance Sheet

2. Present value of contributors liabilities:

$$V_a(t) = V_a^1(t) - V_a^2(t)$$

= discounted value of future benefits of members active at time t

− discounted value of the future contributions of these active members.

First, let us calculate the discounted value of future benefits:

$$V_a^1(t) = \sum_{x=x_0}^{x_r-1} R(t)(1+g)^{x_r-x} \left(\sum_{s=x_r-x}^{\omega-x} L(x+s, t+s) \frac{(1+g)^{s-(x_r-x)}}{(1+r)^s} \right)$$

$$= R(t) \sum_{x=x_0}^{x_r-1} L(x, t) \left(\sum_{s=x_r-x}^{\omega-x} p(x, x+s) \frac{1}{(1+d)^s} \right)$$

$$= R(t) \sum_{x=x_0}^{x_r-1} L(x, t) \frac{1}{(1+d)^{x_r-x}} \ddot{a}_{x_r}.$$

Next, let us calculate the discounted value of future contributions:

$$V_a^2(t) = \pi \sum_{x=x_0}^{x_r-1} \sum_{s=0}^{x_r-x-1} L(x+s, t+s) S(x+s, t+s) \frac{1}{(1+r)^s}$$

$$= \pi S_f(t) \sum_{x=x_0}^{x_r-1} L(x, t) \left(\sum_{s=0}^{x_r-x-1} \Psi(x+s) \frac{1}{(1+d)^s} \right).$$

The provision of active members is therefore given in total by

$$V_a(t) = R(t) \sum_{x=x_0}^{x_r-1} L(x, t) \frac{1}{(1+d)^{x_r-x}} \ddot{a}_{x_r}$$

$$- \pi S_f(t) \sum_{x=x_0}^{x_r-1} L(x, t) \left(\sum_{s=0}^{x_r-x-1} \Psi(x+s) \frac{1}{(1+d)^s} \right)$$

$$= \sum_{x=x_0}^{x_r-1} L(x, t) \left\{ R(t) \ddot{a}_{x_r} \frac{1}{(1+d)^{x_r-x}} \right.$$

$$\left. - \pi S_f(t) \sum_{s=0}^{x_r-x-1} \Psi(x+s) \frac{1}{(1+d)^s} \right\}$$

$$= \sum_{x=x_0}^{x_r-1} L(x, t) NC(x, t),$$

where $NC(x, t)$ is the notional account of an active member of age x at time t.

3. Total provision:

$$V(t) = R(t)\left(\sum_{x=x_r}^{\omega} L(x,t)\ddot{a}_x + \sum_{x=x_0}^{x_r-1} L(x,t)\frac{1}{(1+d)^{x_r-x}}\ddot{a}_{x_r}\right)$$

$$- \pi S_f(t) \sum_{x=x_0}^{x_r-1} L(x,t)\left(\sum_{s=0}^{x-x-1} \Psi(x+s)\frac{1}{(1+d)^s}\right)$$

$$= \sum_{x=x_0}^{\infty} L(x,t)NC(x,t).$$

6.9.3 Contribution Asset and Turnover Duration

In the stationary case, the contribution asset must correspond to the value of the provisions. Unlike funding, there are no financial assets to cover these liabilities. In the logic of Pay-As-You-Go, it seems natural to consider that the wealth of the scheme is based on future contributions. Indeed, by definition of Pay-As-You-Go, the financing of the pension promises, in particular with respect to current retirees, will be achieved by the future contributions of the active population; the latter therefore serve to cover future liabilities.

We therefore propose to express the contribution asset, denoted CA, as a multiple $M(t)$ of the total contributions paid that year, denoted $C(t)$:

$$CA(t) = M(t)C(t).$$

So let us calculate the ratio between the overall provision and the sum of the contributions made in a year.

$$M(t) = \frac{V(t)}{C(t)} = \frac{V_r(t)}{C(t)} + \frac{V_a(t)}{C(t)} = t_r(t) + t_a(t).$$

The total amount of contributions made at time t is given by

$$C(t) = \pi \sum_{x=x_0}^{x_r-1} L(x,t)S(x,t) = \pi S_f(t) \sum_{x=x_0}^{x_r-1} L(x,t)\Psi(x).$$

6.9 Implicit Debt and Pay-As-You-Go Actuarial Balance Sheet

By actuarial balance, the contributions also correspond to the benefits paid in one year:

$$C(t) = R(t) \sum_{x=x_r}^{\omega} L(x,t).$$

We then obtain

1. For calculating the ratio for retirees:

$$t_r(t) = \frac{V_r(t)}{C(t)} = \frac{R(t) \sum_{x=x_r}^{\omega} L(x,t) \ddot{a}_x}{R(t) \sum_{x=x_r}^{\omega} L(x,t)}$$

$$= \frac{\sum_{x=x_r}^{\omega} L(x,t) \left(\sum_{j=x}^{\omega} \frac{L(j,t+j-x)}{L(x,t)} \frac{1}{(1+d)^{j-x}} \right)}{\sum_{x=x_r}^{\omega} L(x,t)}$$

$$= \frac{\sum_{x=x_r}^{\omega} \sum_{r=x}^{\omega} L(j, t+j-x) \frac{1}{(1+d)^{j-x}}}{\sum_{x=x_r}^{o} L(x,t)}$$

$$= \frac{\sum_{x=x_r}^{\omega} \sum_{j=x}^{\omega} L(j,t)}{\sum_{x=x_r}^{\omega} L(x,t)} = \frac{\sum_{x=x_r}^{\omega} L(x,t)(x - x_r + 1)}{\sum_{x=x_r}^{\omega} L(x,t)}$$

$$= \frac{\sum_{x=x_r}^{\omega} L(x,t)x}{\sum_{x=x_r}^{\omega} L(x,t)} - x_r + 1.$$

The term

$$\frac{\sum_{x=x_r}^{\omega} L(x,t)x}{\sum_{x=x_r}^{\omega} L(x,t)}$$

represents the average age of retirees. The ratio for retirees can therefore be interpreted as being equal for a member to the average number of retirement annuities paid from his or her retirement age.

2. For calculating the ratio for active contributors:

$$t_a(t) = \frac{V_a(t)}{C(t)} = \frac{V_a^1(t) - V_a^2(t)}{C(t)}.$$

We have successively:

$$\frac{V_a^1(t)}{C(t)} = \frac{R(t) \sum_{x=x_0}^{x_r-1} L(x,t) \frac{1}{(1+d)^{x_r-x}} \ddot{a}_{x_r}}{R(t) \sum_{x=x_r}^{\omega} L(x,t)}$$

$$= \frac{\sum_{x=x_0}^{x_r-1} L(x,t) \frac{1}{(1+d)^{x_r-x}} \sum_{j=x_r}^{\omega} \frac{L(j,t+j-x)}{L(x_r,t+x_r-x)} \frac{1}{(1+d)^{j-x_r}}}{\sum_{x=x_r}^{\omega} L(x,t)}$$

$$= \frac{\sum_{x=x_0}^{x_r-1} \sum_{j=x_r}^{\omega} L(j,t+j-x) \frac{1}{(1+d)^{j-x}}}{\sum_{x=x_r}^{\omega} L(x,t)} = \frac{\sum_{x=x_0}^{x_r-1} \sum_{j=x_r}^{x} L(j,t)}{\sum_{x=x_r}^{\omega} L(x,t)}$$

$$= x_r - x_0,$$

$$\frac{V_a^2(t)}{C(t)} = \frac{\pi S_f(t) \sum_{x=x_0}^{x_r-1} L(x,t) \left(\sum_{s=0}^{x_r-x-1} \Psi(x+s) \frac{1}{(1+d)^s} \right)}{\pi S(t) \sum_{x=x_0}^{x_r-1} L(x,t) \Psi(x)}$$

$$= \frac{\sum_{x=x_0}^{x_r-1} \sum_{s=0}^{x_r-x-1} L(x+s,t) \Psi(x+s)}{\sum_{x=x_0}^{x_r-1} L(x,t) \Psi(x)}$$

$$= \frac{\sum_{y=x_0}^{x_r-1} \sum_{s=0}^{y-x_0} L(y,t) \Psi(y)}{\sum_{x=x_0}^{x-1} L(x,t) \Psi(x)} = \frac{\sum_{x=x_0}^{x_r-1} (x - x_0 + 1) L(x,t) \Psi(x)}{\sum_{x=x_0}^{x_r-1} L(x,t) \Psi(x)}.$$

The ratio for contributors is therefore given by

$$t_a(t) = \frac{\sum_{x=x_0}^{x_r-1} (x_r - x - 1) L(x,t) \Psi(x)}{\sum_{x=x_0}^{x_r-1} L(x,t) \Psi(x)}.$$

The term

$$\frac{\sum_{x=x_0}^{x_r-1} x L(x,t) \Psi(x)}{\sum_{x=x_0}^{x_0-1} L(x,t) \Psi(x)}$$

represents the average age of active contributors, weighted by the level of salaries. The ratio for contributors can therefore be interpreted as being equal to the weighted average number of contributions made during the active career of each member.

Finally, the ratio of provisions to contributions can be written as

$$M(t) = t_r(t) + t_a(t) = \frac{\sum_{x=x_r}^{\omega} L(x,t) x}{\sum_{x=x_r}^{\omega} L(x,t)} - \frac{\sum_{x=x_0}^{x_0-1} x L(x,t) \Psi(x)}{\sum_{x=x_0}^{x_r-1} L(x,t) \Psi(x)} = TD(t).$$

6.9 Implicit Debt and Pay-As-You-Go Actuarial Balance Sheet

This coefficient, denoted TD, is called turnover duration: it is equal to the difference between the average age of benefit receipt and the average age of contribution payment. It can be interpreted as the average length of time per member that contributions are held before being paid out.

This allows us to define the contribution asset as the product of the total contributions of a year and the turnover duration:

$$CA(t) = TD(t)C(t).$$

The contribution asset therefore corresponds to a "virtual asset" of Pay-As-You-Go to be added in the actuarial balance sheet to establish an actuarial balance sheet of the scheme in a closed system.

The actuarial equilibrium of a Pay-As-You-Go scheme can then be verified in a closed system and automatic adjustment mechanisms can be set up on this basis if necessary.

Chapter 7
Hybrid DB/DC Systems

Abstract Traditional pension schemes, whether social security systems or occupational pension schemes, generally follow one of two design logics: either a Defined Contribution (DC) scheme, where the level of contributions to be paid by the generation of active workers is explicitly defined; retirement benefits are derived by actuarial equivalence; or a Defined Benefit (DB) scheme, where the level of retirement benefits to be paid is explicitly defined; the level of contributions is also derived by actuarial equivalence.

Defined Benefit schemes can be considered as mechanisms where the risks are essentially borne by the contributors to the scheme; in contrast, in Defined Contribution schemes, the risks are shifted to the beneficiaries of the scheme.

To avoid this dichotomy, hybrid schemes have been developed, with an intermediate logic between Defined Benefit and Defined Contribution, in order to better share the risks between the parties.

The objective of this chapter is to study this hybrid scheme logic in a Pay-As-You-Go environment. In particular, we present the Musgrave rule, which is intended to be an intermediate between DB and DC and which aims to guarantee a constant benefit ratio net of contributions. Other intermediate schemes between DB and DC will also be presented. In the case of social security schemes, the ultimate aim is to share risks between generations (working people and retirees) in order to achieve greater inter-generational equity.

7.1 Musgrave Rule

We consider a pure Pay-As-You-Go scheme and adopt the following notation: π is the contribution rate, S is the average salary of working people, R is the average pension of retirees, N_A is the number of active contributors, and N_R is the number of retirees.

The actuarial equilibrium relationship of the Pay-As-You-Go can be written as an equality between the aggregates of contributions and benefits:

$$\pi S N_A = N_R R.$$

Consider an initial balanced situation (denoted by State 1) satisfying the relation:

$$\pi_1 = D_1 \delta_1, \tag{7.1}$$

where $D_1 = N_R/N_A$ is the dependency ratio and $\delta_1 = R/S$ is the benefit ratio of state 1. Let us now suppose that the system evolves towards a new state, denoted state 2, characterized by another dependency ratio denoted D_2. In the context of an aging population, it is normal to assume that $D_2 > D_1$. Given this risk affecting the pension scheme, and therefore its equilibrium, it is appropriate to determine new values for the benefit ratio and the contribution rate, respectively, satisfying the general equilibrium relation:

$$\pi_2 = D_2 \delta_2. \tag{7.2}$$

This equation contains two degrees of freedom to choose from (δ_2 and π_2). There is therefore an infinite number of admissible solutions. It is of course interesting to link the considered solution to the initial pair δ_1 and π_1. In this model, we will call an "automatic adjustment rule" a mechanism which defines the new benefit ratio/contribution rate pair (δ_2, π_2) from the initial pair (δ_1, π_1) and the successive values of the dependency ratio D_1 and D_2. A first simple solution consists in keeping one of the two parameters constant and determining the other parameter in order to satisfy the relation (7.2); in this way, the well-known mechanisms of DB (Defined Benefit) and DC (Defined Contribution) are generated:

1. In a Defined Benefit (DB) scheme, the benefit ratio must remain constant; the adjustment therefore relates entirely to the contribution rate (risk borne only by the active contributors). We then have the following two constraints given the relations (7.1) and (7.2)

$$\pi_2 = D_2 \delta_2,$$

$$\delta_2 = \delta_1 = \frac{\pi_1}{D_1},$$

whose solution is

$$\delta_2 = \delta_1 = \delta,$$

$$\pi_2 = \pi_1 \frac{D_2}{D_1}.$$

Generally, we will have at any state n

$$\delta_n = \delta_{n-1} = \delta,$$

$$\pi_n = \pi_{n-1} \frac{D_n}{D_{n-1}}.$$

7.1 Musgrave Rule

In particular, when the dependency ratio increases, the contribution rate increases in the same proportion.

2. In a Defined Contribution (DC) scheme, the contribution rate must remain constant; the adjustment is therefore entirely on the benefit ratio (risk borne only by the beneficiaries). We then have the following two constraints given relations (7.1) and (7.2):

$$\pi_2 = D_2 \delta_2,$$
$$\pi_2 = \pi_1 = D_1 \delta_1,$$

whose solution is

$$\pi_2 = \pi_1 = \pi,$$
$$\delta_2 = \delta_1 \frac{D_1}{D_2}.$$

Generally, we will have at any state n

$$\pi_n = \pi_{n-1} = \pi,$$
$$\delta_n = \delta_{n-1} \frac{D_{n-1}}{D_n}.$$

In particular, as the dependency ratio increases, the benefit ratio decreases in the same proportion.

In addition to these two adjustment policies, there are an infinite number of other methods that will simultaneously change the value of the benefit ratio and the contribution rate. Musgrave [63] proposed another invariant than the benefit ratio or the contribution rate, leading to a form of risk sharing between working people and pensioners. The Musgrave ratio, denoted M, is defined as the ratio between the average pension and the average wage net of pension contributions. The "gross" benefit ratio δ then becomes a kind of net benefit ratio given by the expression

$$M = \frac{R}{S(1 - \pi)} = \frac{\delta}{(1 - \pi)}. \tag{7.3}$$

It expresses a purchasing power ratio, taking into account that part of the salary of working people is intended to finance pensions and cannot be distributed to working people (unlike retirees who are generally not subject to the pension contribution rate).

In a DB design, the Musgrave ratio increases as the dependency ratio increases:

$$M_2 = \frac{\delta_2}{(1 - \pi_2)} = \frac{\delta_1}{(1 - \pi_1 D_2/D_1)}.$$

One might therefore consider that such a system favors retirees at the expense of working people in terms of purchasing power.

In a DC design, the Musgrave ratio decreases as the dependency ratio increases:

$$M_2 = \frac{\delta_2}{(1-\pi_2)} = \frac{\delta_1}{(1-\pi_1)} \frac{D_1}{D_2}.$$

This time, it could be considered that such a system disadvantages pensioners in terms of purchasing power to the benefit of working people.

The Musgrave rule aims to stabilize this M ratio in the event of a change in the dependency ratio (maintaining the purchasing power ratio of retirees and working people):

$$M_1 = M_2 = M.$$

This type of scheme will be referred to hereafter as "Defined Musgrave" (DM). We can then easily obtain the evolution of the contribution rate and the benefit ratio in such a plan. Given (7.2) and the relationship

$$M = \frac{\delta_2}{(1-\pi_2)} = \frac{\delta_2}{(1-D_2\delta_2)},$$

we obtain successively

$$\delta_2 = \frac{M}{1+MD_2},$$

$$\pi_2 = \frac{D_2 M}{1+MD_2}.$$

We can also compare the evolution of these two quantities from one period to another:

1. Benefit ratio recurrence formula:

$$\delta_2 = \frac{M}{1+MD_2} = \frac{\delta_1/(1-\pi_1)}{1+\frac{\delta_1}{1-\pi_1}D_2}$$

$$= \delta_1 \frac{1}{1+\delta_1(D_2-D_1)}.$$

In a DC plan, the new benefit ratio is given by

$$\delta_2 = \delta_1 \frac{D_1}{D_2}.$$

7.1 Musgrave Rule

In a DB plan, the benefit ratio remains constant by definition:

$$\delta_2 = \delta_1.$$

Let us show that a Defined Musgrave (DM) plan can be considered in terms of the evolution of the benefit ratio as an intermediate between Defined Benefit (DB) and Defined Contribution (DC).

Proposition 7.1 *If the dependency ratio increases* $(D_2 > D_1)$ *and if the initial contribution rate satisfies the natural condition* $0 < \pi_1 < 1$, *we have the following inequality in terms of benefit ratio:*

$$\delta_1 > \delta_1 \frac{1}{1 + \delta_1 (D_2 - D_1)} > \delta_1 \frac{D_1}{D_2} \qquad (7.4)$$

or

$$\delta_2^{DB} > \delta_2^{DM} > \delta_2^{DC}.$$

Proof

a. The inequality

$$\delta_1 > \delta_1 \frac{1}{1 + \delta_1 (D_2 - D_1)}$$

is a direct consequence of the condition $D_2 > D_1$.

b. The inequality

$$\delta_1 \frac{1}{1 + \delta_1 (D_2 - D_1)} > \delta_1 \frac{D_1}{D_2}$$

is equivalent to

$$D_2 > D_1 (1 + \delta_1 (D_2 - D_1)),$$

or to

$$D_2 - D_1 > D_1 \delta_1 (D_2 - D_1) = \pi_1 (D_2 - D_1)$$

or $\pi_1 < 1$.

□

2. Contribution rate recurrence formula:

$$\pi_2 = \frac{D_2 M}{1+MD_2} = \frac{D_2 \frac{\pi_1/D_1}{1-\pi_1}}{1+D_2 \frac{\pi_1/D_1}{1-\pi_1}}$$

$$= \pi_1 \frac{D_2}{D_1 + \pi_1(D_2 - D_1)}.$$

In a DB plan, the new contribution rate is given by

$$\pi_2 = \pi_1 \frac{D_2}{D_1}.$$

In a DC plan, the contribution rate remains constant by definition:

$$\pi_2 = \pi_1.$$

We have the following property for the contribution rate analogous to Proposition 7.1:

Proposition 7.2 *If the dependency ratio increases $(D_2 > D_1)$ and the initial contribution rate satisfies the condition $0 < \pi_1 < 1$, we have the following inequality in terms of contribution rate:*

$$\pi_1 < \pi_1 \frac{D_2}{D_1 + \pi_1(D_2 - D_1)} < \pi_1 \frac{D_2}{D_1} \qquad (7.5)$$

or

$$\pi_1^{DC} = \pi_2^{DC} < \pi_2^{DM} < \pi_2^{DB}.$$

Proof

a. The inequality

$$\pi_1 < \pi_1 \frac{D_2}{D_1 + \pi_1(D_2 - D_1)}$$

is equivalent to $D_1 + \pi_1(D_2 - D_1) < D_2$ or $\pi_1 < 1$.

b. The inequality

$$\pi_1 \frac{D_2}{D_1 + \pi_1(D_2 - D_1)} < \pi_1 \frac{D_2}{D_1}$$

is a direct consequence of the relationship $D_1 + \pi_1(D_2 - D_1) > D_1$. □

7.1 Musgrave Rule

Table 7.1 Contribution rate and benefit ratio formulas in DB, DC and DM

	Contribution rate	Benefit ratio
Defined Benefit	$\pi_n = \pi_{n-1} \frac{D_n}{D_{n-1}}$	$\delta_n = \delta_{n-1} = \delta$
Defined Contribution	$\pi_n = \pi_{n-1} = \pi$	$\delta_n = \delta_{n-1} \frac{D_{n-1}}{D_n}$
Defined Musgrave	$\pi_n = \pi_{n-1} \frac{D_n}{D_{n-1} + \pi_{n-1}(D_n - D_{n-1})}$	$\delta_n = \delta_{n-1} \frac{1}{1 + \delta_{n-1}(D_n - D_{n-1})}$

Table 7.2 Replacement rate in DB, DC and DM according to the dependency ratio

		D_2					
		0.25	0.35	0.40	0.45	0.50	0.60
DB	δ	0.50	0.50	0.50	0.50	0.50	0.50
DC	δ	0.80	0.57	0.50	0.44	0.40	0.33
DM	δ	0.54	0.51	0.50	0.49	0.48	0.45

Table 7.3 Contribution rates in DB, DC and DM according to the dependency ratio

		D_2					
		0.25	0.35	0.40	0.45	0.50	0.60
DB	π	0.13	0.18	0.2	0.23	0.25	0.30
DC	π	0.20	0.20	0.2	0.20	0.20	0.20
DM	π	0.14	0.18	0.2	0.22	0.24	0.27

Formulas (7.4) and (7.5) show that the Musgrave rule can be seen as an intermediate between pure DB and DC in terms of contribution rate and benefit ratio. In a DM scheme, in the event of a demographic shock, these two rates move in opposite directions. Both the contributors to the plan and the beneficiaries are therefore affected. If the dependency ratio rises (aging of the population), there will be a simultaneous increase in the contribution rate and a decrease in the benefit ratio. The opposite phenomenon will occur if the dependency ratio falls.

In general, Table 7.1 below gives the general recurrence formulas for contribution rate and benefit ratio for the three systems (DB, DC and DM) at any state n.

Example 7.1 Consider an initial steady state characterized by the following parameters: dependency ratio: $D = 0.40$; benefit ratio: $\delta = 0.50$; contribution rate: $\pi = 0.40 \times 0.50 = 0.20$. We assume that the dependency ratio changes and we look at the effect on the contribution rate and benefit ratio in the three formulas DB, DC, DM. Tables 7.2 and 7.3 give the corresponding benefit ratio and contribution rate.

7.2 General Risk Sharing Mechanisms

Unlike the classic Defined Benefit of Defined Contribution formulas, the Musgrave rule presented above makes it possible to share the demographic risk between the two generations (active and retired). But this is of course only one particular formula. Various families of intermediate formulas between DB and DC can be developed. Each formula differs from the others in the level of risk sharing between working people and retirees. In this context, DB and DC appear to be two extremes, one in which the entire risk is borne by working people, the other by retirees.

7.2.1 Proportional Risk Sharing

In order to introduce a first family of intermediate formulas, we start from an initial state satisfying the equilibrium relation (7.1).

A change in the dependency ratio D will generate new values for the contribution rate and the benefit ratio, always satisfying the equilibrium relationship (7.2). In order to highlight the successive variations in contribution and benefit ratio, let us introduce the following notation:

$$\begin{aligned} \pi_2 &= \pi_1 \left(1 + \lambda_\pi\right), \\ \delta_2 &= \delta_1 \left(1 - \lambda_\delta\right). \end{aligned} \quad (7.6)$$

The parameters λ_π and λ_δ thus represent rates of change. The notation chosen suggests, as already illustrated, that the contribution rate and benefit ratio are likely to move in opposite directions (in order to respect Relationship (7.2)).

Relations (7.1) and (7.2) allow us to obtain a link between these parameters:

$$(1 + \lambda_\pi) = \frac{D_2}{D_1} (1 - \lambda_\delta). \quad (7.7)$$

We can also define a risk-sharing coefficient that compares the efforts borne by working people and retirees:

$$\mu = \frac{\lambda_\pi}{\lambda_\delta}. \quad (7.8)$$

Let us first calculate the values of this coefficient in the three formulas highlighted in Sect. 7.1 (DB, DC, DM).

1. In a Defined Benefit (DB) plan, we have by definition:

$$\lambda_\delta = 0.$$

7.2 General Risk Sharing Mechanisms

Equation (7.7) then gives the other parameter:

$$1 + \lambda_\pi = \frac{D_2}{D_1} \quad \Leftrightarrow \quad \lambda_\pi = \frac{D_2 - D_1}{D_1}.$$

The risk-sharing coefficient becomes

$$\mu = +\infty.$$

2. In a Defined Contribution (DC) design, we have by definition

$$\lambda_\pi = 0.$$

Equation (7.7) then gives the other parameter:

$$1 = \frac{D_2}{D_1}(1 - \lambda_\delta) \quad \Leftrightarrow \quad \lambda_\delta = \frac{D_2 - D_1}{D_2}.$$

The risk-sharing coefficient becomes

$$\mu = 0.$$

3. In a Defined Musgrave scheme, using Table 7.1, we get for the contribution rate

$$\pi_2 = \pi_1 \frac{D_2}{D_1 + \pi_1(D_2 - D_1)} = \pi_1(1 + \lambda_\pi) \quad \Leftrightarrow \quad \lambda_\pi = \frac{(D_2 - D_1)(1 - \pi_1)}{D_1 + \pi_1(D_2 - D_1)}$$

and for the benefit ratio

$$\delta_2 = \delta_1 \frac{1}{1 + \delta_1(D_2 - D_1)} = \delta_1(1 - \lambda_\delta) \quad \Leftrightarrow \quad \lambda_\delta = \frac{\delta_1(D_2 - D_1)}{1 + \delta_1(D_2 - D_1)}.$$

The risk-sharing coefficient thus becomes

$$\mu = \frac{\lambda_\pi}{\lambda_\delta} = \frac{(D_2 - D_1)(1 - \pi_1)}{D_1 + \pi_1(D_2 - D_1)} \frac{1 + \delta_1(D_2 - D_1)}{\delta_1(D_2 - D_1)}$$

$$= \frac{1 - \pi_1}{\pi_1} = \frac{1}{\pi_1} - 1.$$

For usual values of the contribution rate ($\pi < 0.5$), this risk-sharing coefficient is greater than 1, suggesting that by accepting this measure, contributors deliver a greater proportional effort in DM than do retirees. This DM system can therefore be considered closer to a pure DB system than to a pure DC system.

As an illustration, in Example 7.1 ($\pi_1 = 0.2$), the risk-sharing coefficient is equal to 4.

Note that in the Defined Musgrave formula, this risk-sharing coefficient does not remain constant but depends on the contribution rate, which itself varies in this philosophy. As a result, in a multi-period model with successive annual applications of the Musgrave rule, the risk-sharing coefficient will change over time. We will then have at time n

$$\mu_n = \frac{1}{\pi_{n-1}} - 1.$$

In addition to the three basic formulas developed above, we can consider other forms of risk sharing.

4. As an example, a natural candidate would be to have a system with a risk-sharing coefficient always equal to 1. We could call this system DE (Defined Equal sharing).

   In such a case, we have

   $$\lambda_\pi = \lambda_\delta = \lambda.$$

Equation (7.7) then becomes

$$(1 + \lambda) = \frac{D_2}{D_1}(1 - \lambda) \quad \Leftrightarrow \quad \lambda = \frac{D_2 - D_1}{D_2 + D_1}.$$

Given Relationship (7.6), the new contribution rate and benefit ratio become

$$\pi_2 = \pi_1 \frac{2D_2}{D_1 + D_2},$$

$$\delta_2 = \delta_1 \frac{2D_1}{D_1 + D_2}.$$

5. Generally speaking, a pension scheme can be characterized by its degree of solidarity between active and retired members through its risk-sharing coefficient. On this basis, we can consider that a value of $\mu > 1$ (resp. $\mu < 1$) generates more effort on the part of contributors (resp. retirees), the classical formulas DB and DC being the limit cases.

   In a formula where this coefficient μ is constant, we have the two relations

   $$(1 + \lambda_\pi) = \frac{D_2}{D_1}(1 - \lambda_\delta),$$

   $$\lambda_\pi = \mu \lambda_\delta.$$

7.2 General Risk Sharing Mechanisms

The solution in terms of λ is given by

$$\lambda_\delta(\mu) = \frac{D_2 - D_1}{D_2 + \mu D_1},$$

$$\lambda_\pi(\mu) = \mu \frac{D_2 - D_1}{D_2 + \mu D_1}.$$

The contribution rate and benefit ratio become

$$\pi_2 = \pi_1 \frac{D_2(1+\mu)}{D_2 + \mu D_1},$$

$$\delta_2 = \delta_1 \frac{D_1(1+\mu)}{D_2 + \mu D_1}. \quad (7.9)$$

Generally speaking, at any time n, we have recurrently:

$$\pi_n = \pi_{n-1} \frac{D_n(1+\mu)}{D_n + \mu D_{n-1}},$$

$$\delta_n = \delta_{n-1} \frac{D_{n-1}(1+\mu)}{D_n + \mu D_{n-1}}.$$

Example 7.2 Using the same assumptions as in Example 7.1, Tables 7.4 and 7.5 illustrate the effect of demographic risk (change in dependency ratio D) as a function of the risk-sharing coefficient μ on the benefit ratio and contribution rate, respectively.

Table 7.4 Comparison of the effect of demographic risk on the benefit ratio as a function of the risk-sharing coefficient μ (initial dependency ratio: $D_1 = 0.40$)

μ		D_2				
		0.25	0.35	0.45	0.50	0.60
0	DC	0.8	0.57	0.44	0.40	0.33
0.5		0.67	0.55	0.46	0.43	0.38
1	DE	0.62	0.53	0.47	0.44	0.40
4	DM	0.54	0.51	0.49	0.48	0.45
20		0.51	0.50	0.50	0.49	0.49
∞	DB	0.50	0.50	0.50	0.50	0.50

Table 7.5 Comparison of the effect of demographic risk on the contribution rate as a function of the risk-sharing coefficient μ (initial dependency ratio: D_1=0.40)

μ		D_2				
		0.25	0.35	0.45	0.50	0.60
0	DC	0.20	0.20	0.20	0.20	0.20
0.5		0.17	0.19	0.21	0.21	0.23
1	DE	0.15	0.19	0.21	0.22	0.24
4	DM	0.14	0.18	0.22	0.24	0.27
20		0.13	0.18	0.22	0.25	0.29
∞	DB	0.13	0.18	0.23	0.25	0.30

7.2.2 Convex Risk Sharing

In addition to the risk-sharing coefficient, which can characterize a plan's philosophy of adaptation to demographic risk, the degree of solidarity between working people and retirees can also be measured using another coefficient, motivated by an interpretation of the Musgrave rule.

Indeed, the Musgrave rule can be expressed as a convex linear combination of contribution rate and benefit ratio, which must remain constant over time. This alternative interpretation of Musgrave rule will allow us to generate a whole family of intermediate formulas between DB and DC.

The DM formula is based on the following invariant (cf. (7.3)):

$$M = \frac{\delta}{1 - \pi}.$$

This relationship can be written as

$$M = \delta + M\pi \quad \Leftrightarrow \quad \frac{1}{1+M}\delta + \frac{M}{1+M}\pi = \frac{M}{1+M}.$$

This in turn can be expressed as a constant convex linear combination of benefit ratio and contribution rate:

$$\eta_M \delta + (1 - \eta_M)\pi = 1 - \eta_M, \tag{7.10}$$

with

$$0 \leq \eta_M = \frac{1}{1+M} \leq 1.$$

7.2 General Risk Sharing Mechanisms

The relation (7.10) allows us to generalize by considering other possible convex invariants.

In general, any coefficient $0 \leq \eta \leq 1$ can generate a risk sharing formula such that

$$\eta \delta + (1-\eta)\pi = \text{constant}. \tag{7.11}$$

We can already easily observe that a DB plan corresponds to the case $\eta = 1$ and a DC plan to the case $\eta = 0$. This coefficient can be interpreted as a measure of solidarity between working people and retirees. A low value of η (close to 0) indicates that the weight of the adjustment falls mainly on retirees (and even totally if $\eta = 0$ in DC). Conversely, a high value of η (close to 1) suggests that the burden of the adjustment falls mainly on working people (and even totally if $\eta = 1$ in DB).

In the event of a demographic shock, the new contribution rate and benefit ratio then obey the following equations:

$$\eta \delta_2 + (1-\eta)\pi_2 = \eta \delta_1 + (1-\eta)\pi_1,$$
$$\pi_2 = D_2 \delta_2,$$

whose solution is

$$\delta_2 = \delta_1 \frac{\eta + (1-\eta)D_1}{\eta + (1-\eta)D_2},$$
$$\pi_2 = \pi_1 \frac{D_2}{D_1} \frac{\eta + (1-\eta)D_1}{\eta + (1-\eta)D_2}. \tag{7.12}$$

In general, we will have at any time n

$$\delta_n = \delta_{n-1} \frac{\eta + (1-\eta)D_{n-1}}{\eta + (1-\eta)D_n},$$
$$\pi_n = \pi_{n-1} \frac{D_n}{D_{n-1}} \frac{\eta + (1-\eta)D_{n-1}}{\eta + (1-\eta)D_n}.$$

For example, taking $\eta = 0.5$ (equal weight between contribution rate and benefit ratio), we have

$$\delta_2 = \delta_1 \frac{1+D_1}{1+D_2},$$
$$\pi_2 = \pi_1 \frac{D_2}{D_1} \frac{1+D_1}{1+D_2}.$$

Table 7.6 Contribution rate and benefit ratio obtained by varying the parameter η

Coefficient		Contribution rate	Benefit ratio
0%	DC	0.20	0.40
25%		0.22	0.44
50%		0.23	0.47
61.54%	DM	0.24	0.48
100%	DB	0.25	0.50

Example 7.3 Using the same assumptions as in Example 7.1, and for a dependency ratio that increases from an initial value of 40% to a value of 50%, Table 7.6 illustrates the effect of demographic risk (change in the dependency ratio D) on the benefit ratio and the contribution rate (formulas (7.12)). In this example, we can calculate the value of the convex coefficient corresponding to the Musgrave rule. The Musgrave invariant is given in this example by

$$M = \frac{\delta}{1-\pi} = \frac{0.5}{1-0.20} = 0.625.$$

The corresponding coefficient η is therefore equal to

$$\eta = \frac{1}{1+M} = \frac{1}{1.625} = 0.6154.$$

The following proposition allows us to relate this convex coefficient η to the risk-sharing coefficient μ introduced in (7.8).

Proposition 7.3 *The risk-sharing coefficient given by (7.8) and the convex parameter given by (7.11) are related by the relationship*

$$\mu = \frac{1}{D} \frac{\eta}{1-\eta}. \tag{7.13}$$

Proof The benefit ratio, expressed as a function of the coefficient μ, is given by (7.9)

$$\delta_2 = \delta_1 \frac{D_1(1+\mu)}{D_2 + \mu D_1}.$$

On the other hand, the benefit ratio, expressed as a function of the coefficient η is given by (7.12)

$$\delta_2 = \delta_1 \frac{\eta + (1-\eta)D_1}{\eta + (1-\eta)D_2}.$$

7.3 Continuous Time Model

So we have

$$\frac{\eta + (1-\eta)D_1}{\eta + (1-\eta)D_2} = \frac{D_1(1+\mu)}{D_2 + \mu D_1},$$

which indeed gives after elementary calculation

$$\mu = \frac{1}{D_1}\frac{\eta}{1-\eta}.$$

It will be noticed that the link expressed by the relation (7.13) is dynamic in time in connection with the evolution in time of the dependency ratio D. □

7.3 Continuous Time Model

The models presented in Sects. 7.1 and 7.2 were developed in discrete time. We can also consider a continuous time modeling which will allow us to highlight the differential equations of the parameters and to understand the local dynamics.

In this context, generalizing the notation used previously, for a time index $t \in [0, T] \subset \mathbb{R}$: $S(t)$ is the average wage at time t, $R(t)$ is the average pension at time t, $N_A(t)$ is the number of active contributors at time t, $N_R(t)$ is the number of retired beneficiaries at time t, and $\pi(t)$ is the contribution rate at time t.

We will also use the average benefit ratio and the dependency ratio, given respectively by

$$\delta(t) = \frac{R(t)}{S(t)},$$

$$D(t) = \frac{N_R(t)}{N_A(t)}.$$

We can then highlight instantaneous revenues and expenses:

$$\text{Revenues}(t) = \pi(t)S(t)N_A(t),$$
$$\text{Expenses}(t) = R(t)N_R(t).$$

The equality between revenues and expenses leads to the equilibrium equation:

$$\pi(t)S(t)N_A(t) = R(t)N_R(t)$$

or

$$\pi(t) = \frac{R(t)}{S(t)}\frac{N_A(t)}{N_R(t)}$$

or

$$\pi(t) = \frac{R(t)}{S(t)} D(t) \qquad (7.14)$$

or

$$\pi(t) = \delta(t) D(t). \qquad (7.15)$$

The equilibrium relationship of the Pay-As-You-Go is the continuous-time equivalent of the discrete relationship. The risk factor in this relationship (independent variable) is the dependency ratio D. Assuming these deterministic time functions are sufficiently regular, we can put these relationships into differential form.

We obtain, by taking the logarithm of Eq. (7.14):

$$\ln \pi(t) = \ln D(t) + \ln R(t) - \ln S(t)$$

and by taking the differential:

$$\frac{d\pi(t)}{\pi(t)} = \frac{dR(t)}{R(t)} - \frac{dS(t)}{S(t)} + \frac{dD(t)}{D(t)}. \qquad (7.16)$$

This relationship expresses that the proportional change in the contribution rate is caused by two effects:

$$\frac{dR(t)}{R(t)} - \frac{dS(t)}{S(t)} = \text{Difference of evolution between pension and salary,}$$

$$\frac{dD(t)}{D(t)} = \text{Aging effect.}$$

A similar relationship can also be shown for the benefit ratio by similarly taking the derivative of the relationship (7.14); we then obtain

$$\frac{d\pi(t)}{\pi(t)} = \frac{d\delta(t)}{\delta(t)} + \frac{dD(t)}{D(t)}, \qquad (7.17)$$

expressing that the change in the contribution rate results from the superposition of the change in the benefit ratio and the dependency ratio.

In this relationship, the change in the dependency ratio D is an exogenous effect on the plan (the effect of aging); as a result of this change, the contribution rate and the benefit ratio must be adjusted to always satisfy the equilibrium relationship (7.16). Different adjustment formulas are possible (one equation with two unknowns).

In this context, the two classic extreme solutions of Defined Benefit (DB) on final salary and Defined Contribution (DC) can be highlighted.

Case 1: Defined Benefit

The benefit ratio remains constant. Relationship (7.17) therefore becomes

$$\frac{d\pi(t)}{\pi(t)} = \frac{dD(t)}{D(t)},$$

expressing that the entire demographic risk is borne by the working population. An increase in the dependency ratio is fully reflected in the form of an increase in the contribution rate.

Relation (7.16) further implies

$$\frac{dR(t)}{R(t)} = \frac{dS(t)}{S(t)}. \tag{7.18}$$

Pensions follow perfectly the evolution of salaries.

Case 2: Defined Contribution

The contribution rate remains constant. Relationship (7.17) therefore becomes

$$\frac{d\delta(t)}{\delta(t)} = -\frac{dD(t)}{D(t)},$$

expressing that the entire demographic risk falls on pensioners. An increase in the dependency ratio is fully reflected in the form of a decrease in the benefit ratio.

Relation (7.16) further implies

$$\frac{dR(t)}{R(t)} = \frac{dS(t)}{S(t)} - \frac{dD(t)}{D(t)}.$$

The rate of change of pensions is fully impacted by the risk of aging.

General Case: Demographic Risk Sharing

Rather than having a total impact on either the contribution rate or the benefit ratio, we can consider risk sharing. The differential effect of aging $\frac{dD(t)}{D(t)}$ will then be distributed according to a proportion (potentially variable over time) $\alpha(t)$ on the benefit ratio and a proportion $1 - \alpha(t)$ on the contribution rate (with $0 \leq \alpha(t) \leq 1$). The relationship (7.17) can indeed be written

$$\frac{d\pi(t)}{\pi(t)} - \frac{d\delta(t)}{\delta(t)} = \frac{dD(t)}{D(t)}.$$

We then distribute the aging effect in such a way as to satisfy the overall effect of aging (7.17):

$$\frac{d\pi(t)}{\pi(t)} = (1 - \alpha(t))\frac{dD(t)}{D(t)},$$
$$-\frac{d\delta(t)}{\delta(t)} = \alpha(t)\frac{dD(t)}{D(t)}.$$
(7.19)

The function α can be interpreted as an automatic adjustment factor for demographic risk, to be chosen according to the desired degree of solidarity between retirees and working people ($\alpha \equiv 0$ in DB; $\alpha \equiv 1$ in DC). In particular, by choosing a constant $0 \leq \alpha \leq 1$, we have for the contribution rate

$$\frac{d\pi(t)}{\pi(t)} = (1 - \alpha)\frac{dD(t)}{D(t)},$$

whose solution is

$$\pi(t) = A\, D(t)^{1-\alpha}, \qquad (7.20)$$

where A is a constant defined by the initial values

$$A = \frac{\pi_0}{D_0^{1-\alpha}}.$$

Taking into account the relation (7.14) we then get directly for the benefit ratio

$$\delta(t) = AD(t)^{-\alpha}. \qquad (7.21)$$

Relationships (7.20) and (7.21) above illustrate the effect of aging on contributions and benefits. In particular, for $\alpha = 0.5$, the same proportion of demographic risk affects the variation in the contribution rate and the benefit ratio.

We can also find in this continuous time formalization the Musgrave rule introduced above. The Musgrave formula is based on the following invariant (cf. (7.3)):

$$M = \frac{R(t)}{S(t)(1 - \pi(t))} = \frac{\delta(t)}{(1 - \pi(t))}.$$

Taking the logarithm and differentiating, we obtain

$$\ln \delta(t) = \ln M + \ln(1 - \pi(t))$$
$$\frac{d\delta(t)}{\delta(t)} = -\frac{d\pi(t)}{1 - \pi(t)}.$$

7.3 Continuous Time Model

Substituting this into (7.17), we get

$$\frac{d\pi(t)}{\pi(t)} = -\frac{d\pi(t)}{1-\pi(t)} + \frac{dD(t)}{D(t)},$$

$$\frac{d\pi(t)}{\pi(t)}\left(1 + \frac{\pi(t)}{1-\pi(t)}\right) = \frac{dD(t)}{D(t)}.$$

Finally,

$$\frac{d\pi(t)}{\pi(t)} = (1-\pi(t))\frac{dD(t)}{D(t)}. \tag{7.22}$$

By analogy with the relationship in (7.19), we deduce that the Musgrave rule corresponds to an adjustment coefficient equal to the contribution rate:

$$\alpha(t) = \pi(t).$$

Applying the Musgrave rule therefore leads to assigning a proportion $\pi(t)$ of the instantaneous change in the dependency ratio to the benefit ratio and the complementary proportion $1 - \pi(t)$ to the contribution rate.

We can obtain an explicit solution of the differential Eq. (7.22).

Proposition 7.4 *The Musgrave formula leads in continuous time to the following forms of the contribution rate and the benefit ratio, K being a constant:*

$$\pi(t) = \frac{KD(t)}{1 + KD(t)},$$

$$\delta(t) = \frac{K}{1 + KD(t)}.$$

Proof The Musgrave differential Eq. (7.22) can be put into logarithmic form:

$$\frac{d\ln \pi(t)}{1 - \pi(t)} = d\ln D(t). \tag{7.23}$$

Defining $Z(t) = \ln \pi(t)$ and $Y(t) = \ln D(t)$, we have

$$\frac{dZ(t)}{1 - e^{Z(t)}} = dY(t).$$

Since

$$\int \frac{dx}{1 - e^x} = x - \ln\left(1 - e^x\right),$$

we get by integrating the two sides of (7.23)

$$\ln \frac{\pi(t)}{1 - \pi(t)} = \ln D(t) + \ln K,$$

and thus finally

$$\pi(t) = \frac{KD(t)}{1 + KD(t)},$$

and by the general equilibrium Eq. (7.15)

$$\delta(t) = \frac{\pi(t)}{D(t)} = \frac{K}{1 + KD(t)},$$

where the constant K depends on the initial conditions:

$$K = \frac{\pi_0}{D_0 (1 - \pi_0)}.$$

□

Example 7.4 The dependency ratio D is assumed to follow a mean-reverting process of the following form:

$$D(t) = D_0 e^{-\beta t} + D^* \left(1 - e^{-\beta t}\right).$$

This relation expresses that the dependency ratio initially starts from the value D_0 and converges asymptotically to the value D^*, the speed of the convergence being controlled by the parameter β. Consider the following assumptions:

- The dependency ratio D is initially equal to $D_0 = 40\%$ and in the growth phase tends towards a limit value $D^* = 66\%$,
- The initial contribution rate is 20%; the initial benefit ratio is therefore 50%.

Figures 7.1 and 7.2 illustrate the evolution over a 40-year horizon, on the one hand of the benefit ratio, and on the other hand of the contribution rate in 4 schemes: DB (Defined Benefit), DC (Defined Contribution), DM (Musgrave rule) and DE (coefficient $\alpha = 0.5$).

7.4 Differentiated Adjustment on Current Pensions

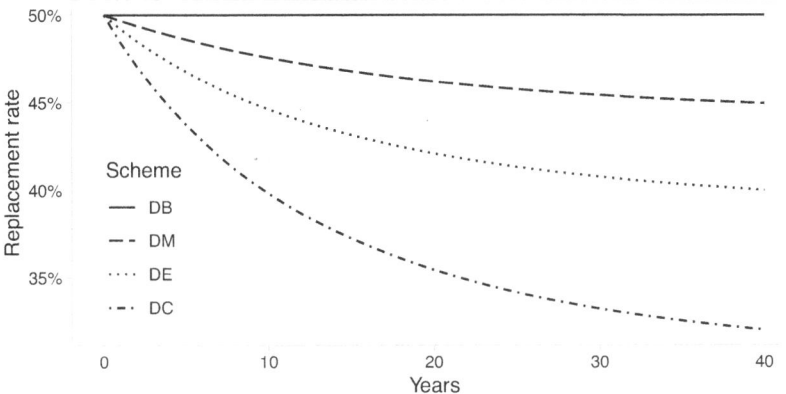

Fig. 7.1 Evolution of the benefit ratio in Example 7.4

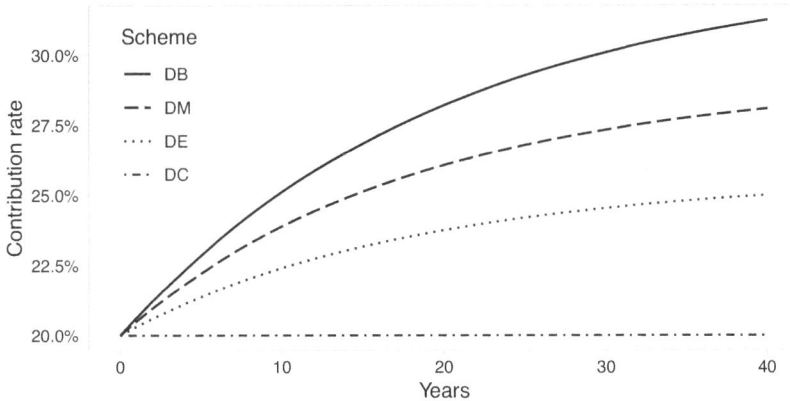

Fig. 7.2 Evolution of the contribution rate in Example 7.4

7.4 Differentiated Adjustment on Current Pensions

The models presented above implicitly assume that the same rules for adjusting average pensions apply to new retirees of the year on the one hand and to already retired members on the other. This is because the same benefit ratio is applied to both new and old retirees. This can lead to a decrease in pensions in payment; in this context, it might be wise to apply different annual adjustment rules to the benefit ratio of new retirees and to the annual revaluation of old pensions. The benefit ratio can then be considered as an average over the whole population of pensioners, resulting from the application of two different automatic adjustment mechanisms.

As an example, consider a discrete time model, characterized each year by two generations of retirees: the new retirees of the year and the old retirees (the generation retired the previous year). Inspired by the previous section, we introduce the following notation: π_n is the contribution rate in year n, $S(n)$ is the average wage of working people, $R(n)$ is the average pension of retirees, $N_A(n)$ is the number of active contributors, $N_R(n)$ is the number of retirees, $N_R^{(1)}(n)$ is the number of new retirees, and $N_R^{(2)}(n)$ is the number of old retirees, so that $N_R(n) = N_R^{(1)}(n) + N_R^{(2)}(n)$.

The average pension $R(n)$ for the whole population then results from an average pension level, on the one hand for the new generation retiring at time n, and on the other hand for the generation that retired the year before:

$$N_R(n)R(n) = N_R^{(1)}(n)R^{(1)}(n) + N_R^{(2)}(n)R^{(2)}(n), \qquad (7.24)$$

with $R^{(1)}(n)$ = average pension for the new generation of retirees and $R^{(2)}(n)$ = average pension for the old generation of retirees.

The average benefit ratio is given by

$$\delta_n = \frac{R(n)}{S(n)}.$$

This average benefit ratio results, on the one hand, from the benefit ratio granted to the new generation of retirees, and on the other hand, from the adjustment of the pension between time $n-1$ and time n for the generation that retired in the previous period:

1. For the new generation of retirees, we will have a benefit ratio given by

$$\delta_n^{(1)} = \frac{R^{(1)}(n)}{S(n)}. \qquad (7.25)$$

2. For the older generation, the pension can be written as a function of the pension granted the year before; if the average pension were adjusted at the same rate as average wages, we would have

$$R^{(2)}(n) = R^{(1)}(n-1)\frac{S(n)}{S(n-1)}. \qquad (7.26)$$

If the average benefit ratio δ should have to decrease, it seems fair to take this into account also in the adjustment of current pensions by keeping the possibility of not adjusting pensions entirely on salaries, but only partially; we then introduce

7.4 Differentiated Adjustment on Current Pensions

a sustainability coefficient, denoted β, less than or equal to one, such that

$$R^{(2)}(n) = R^{(1)}(n-1)\frac{S(n)}{S(n-1)}\beta_n. \tag{7.27}$$

Taking into account the relations (7.24), (7.25) and (7.27), we can link the different parameters of the model:

$$N_R(n)\delta_n = N_R^{(1)}(n)\delta_n^{(1)} + N_R^{(2)}(n)\delta_{n-1}^{(1)}\beta_n.$$

Defining

$$\text{trr}(n) = \frac{N_R^{(1)}(n)}{N_R(n)} = \text{rate of renewal of retirees},$$

we obtain

$$\delta_n = \text{trr}(n)\delta_n^{(1)} + (1 - \text{trr}(n))\delta_{n-1}^{(1)}\beta_n. \tag{7.28}$$

On the other hand, we still have the equilibrium equation in Pay-As-You-Go linking the contribution rate and the global benefit ratio:

$$\pi_n = \delta_n D_n. \tag{7.29}$$

A risk-sharing mechanism can then be characterized as a two-step procedure:

(a) Step 1—distribution of risks between retirees and active workers (intergenerational distribution): choice of the parameters δ and π satisfying the relation (7.29) according to the rules developed in Sect. 7.2 (for example by applying the Musgrave rule),
(b) Step 2—distribution of risks within the retirees (intra-generational distribution): once the global benefit ratio δ is fixed by the first step, we choose the parameters $\delta_n^{(1)}$ and β_n satisfying the relation (7.28).

For each of these two steps, we have one equation (respectively Eqs. (7.29) and (7.28)) and two variables (respectively variables π_n, δ_n and variables $\delta_n^{(1)}, \beta_n$); thus, at each step there are many possible risk sharing mechanisms. We presented in Sect. 7.2 some possible sharing mechanisms for the first stage of allocation between active and retired people.

Let us now detail some possible sharing schemes for the second step within the retirees.

7.4.1 Proportional Risk Sharing Between the two Retirees Generations

In this case, the overall benefit ratio is applied to new retirees at all times:

$$\delta_n^{(1)} = \delta_n.$$

From Eq. (7.28), we obtain a value of the sustainability coefficient given by

$$\beta_n = \frac{\delta_n}{\delta_{n-1}^{(1)}} = \frac{\delta_n}{\delta_{n-1}}. \tag{7.30}$$

Given (7.26), the new pension paid to the previously retired generation then becomes

$$R^{(2)}(n) = R^{(1)}(n-1)\frac{S(n)}{S(n-1)}\frac{\delta_n}{\delta_{n-1}} = \delta_{n-1}S(n-1)\frac{S(n)}{S(n-1)}\frac{\delta_n}{\delta_{n-1}}$$
$$= \delta_n S(n) = R^{(1)}(n).$$

Both generations therefore receive the same pension at the same time. The risk of a possible decrease in the overall benefit ratio affects them both in the same way.

7.4.2 Risk Entirely Borne by Older Retirees

In this case, a fixed benefit ratio is applied to new retirees at all times:

$$\delta_n^{(1)} = \bar{\delta}.$$

The entire cost of any adjustment resulting from a decline in the overall benefit ratio is reflected in the adjustment rate of the old pensions. The equilibrium Eq. (7.28) becomes

$$\delta_n = \text{trr}(n)\bar{\delta} + (1 - \text{trr}(n))\bar{\delta}\beta_n.$$

The sustainability coefficient is then given by

$$\beta_n = \frac{\delta_n - \text{trr}(n)\bar{\delta}}{\bar{\delta} - \text{trr}(n)\bar{\delta}} = 1 - \frac{\bar{\delta} - \delta_n}{\bar{\delta}(1 - \text{trr}(n))}.$$

7.4.3 Risk Entirely Borne by Younger Retirees

In this case, a sustainability coefficient equal to one is applied at all times to former pensioners (complete adaptation of current pensions to salaries):

$$\beta_n = 1.$$

The equilibrium equation becomes

$$\delta_n = \operatorname{trr}(n)\delta_n^{(1)} + (1 - \operatorname{trr}(n))\delta_{n-1}^{(1)}.$$

The benefit ratio of the new pensions is then

$$\delta_n^{(1)} = \frac{\delta_n - (1 - \operatorname{trr}(n))\delta_{n-1}^{(1)}}{\operatorname{trr}(n)}$$

$$= \delta_{n-1}^{(1)} + \frac{\delta_n - \delta_{n-1}^{(1)}}{\operatorname{trr}(n)}.$$

7.4.4 Proportional Risk Sharing with Guarantee

This scenario is a variant of scenario (a) where a ratchet guarantee on pension adjustment is added. Taking into account the presence of the sustainability coefficient, Formula (7.27) shows that the amount of the pension can, for the same pensioner, decrease nominally from one year to the next. If we want to guarantee at least the same pension in year n as that received for this generation in year $n - 1$, we must add the constraint:

$$R^{(2)}(n) = R^{(1)}(n-1)\frac{S(n)}{S(n-1)}\beta_n \geq R^{(1)}(n-1).$$

i.e. the following constraint on the sustainability coefficient:

$$\beta_n \geq \frac{S(n-1)}{S(n)}.$$

In particular, if we consider proportional sharing, the sustainability coefficient (7.30) becomes

$$\beta_n = \max\left(\frac{\delta_n}{\delta_{n-1}^{(1)}}, \frac{S(n-1)}{S(n)}\right).$$

The benefit ratio for the new generation is then

$$\delta_n^{(1)} = \frac{\delta_n - (1 - \text{trr}(n))\delta_{n-1}^{(1)} \max\left(\frac{\delta_n}{\delta_{n-1}^{(1)}}, \frac{S(n-1)}{S(n)}\right)}{\text{trr}(n)}.$$

Other types of guarantees can of course be considered and integrated in a similar way.

Chapter 8
Point Systems

Abstract In this chapter, we develop various actuarial considerations relating to point systems. A point system is a very general methodology which can be seen as a change of currency: rather than accumulating rights in monetary units, one accumulates rights in the form of points. In this framework, one can then work with Defined Contribution as well as with Defined Benefit or with a hybrid formula between DC and DB. In this chapter, we will develop the actuarial analysis of both Defined Contribution point systems and hybrid DB/DC systems.

8.1 Pay-As-You-Go DC Point Systems

We begin by describing a Defined Contribution Pay-As-You-Go point scheme. This type of plan is therefore quite similar to notional accounts, since it combines pure Pay-As-You-Go financing with a Defined Contribution logic.

8.1.1 General Principles

We assume a scheme with contributions expressed as a percentage of the salary taken into account, which we will call the reference salary; this can be the full salary, a capped salary or a salary range between two defined ceilings. Each year, the member's contribution entitles him/her to a certain number of pension points. The number of points is given by the ratio between the contribution paid and the purchase value, which is the same for everyone, called the acquisition price of the point.

Denoting by $S^{(i)}(t)$ the reference wage at time t of individual i, π the fixed contribution rate to the plan, $S_r(t)$ the price of the point at time t (e.g. the average wage in the economy), $n^{(i)}(t)$ the number of points earned at time t for member i,

© The Author(s), under exclusive license to Springer Nature Switzerland AG 2025
P. Devolder, S. de Valeriola, *Actuarial Pension Funding Theory*, Springer Actuarial Textbooks, https://doi.org/10.1007/978-3-031-85268-8_8

x_0 the age of affiliation and x_r the retirement age (which is assumed in this section to be uniform), we have

$$n^{(i)}(t) = \pi \frac{S^{(i)}(t)}{S_r(t)}. \tag{8.1}$$

At retirement age x_r, the member i has thus acquired a total number of points equal to

$$NP^{(i)} = \sum_{t=t_0(i)}^{t_1(i)-1} n^{(i)}(t),$$

where $t_0(i)$ is the time of affiliation of individual i (at age x_0) and $t_1(i)$ is the time of retirement of individual i (at age x_r).

From time $t_1(i)$ a retirement benefit denoted $R^i(t)$ can be paid each year t to the (alive) individual i, according to the formula

$$R^{(i)}(t) = VP(t)NP^{(i)} \quad (t \geq t_1(i)), \tag{8.2}$$

where $VP(t)$ is the value of the point at time t.

The value of the point $VP(t)$ is set in such a way as to respect the principle of actuarial equilibrium at the level of the entire plan (overall equivalence between contributions received and benefits paid). Given the number of parameters, different methods of calculating this value can be proposed, depending in particular on the method of revaluating pensions after retirement age.

8.1.2 Standard Actuarial Equilibrium

In this case, we try to obtain a uniform value of the point for all the retirees. The general actuarial equilibrium relation underlying Pay-As-You-Go is applied at all times: equivalence between contributions received and pensions paid.

The contributions collected at time t are given by

$$\pi \sum_i S^{(i)}(t),$$

where the summation is over all individuals i active at time t and contributing to the plan.

The benefits to be paid at time t are given by

$$VP(t) \sum_j NP^{(j)},$$

8.1 Pay-As-You-Go DC Point Systems

where the summation is over all individuals j retired at time t.

Equating these two aggregates, we obtain a value of the point given by

$$VP(t) = \frac{\pi \sum_i S^{(i)}(t)}{\sum_j NP^{(j)}}.$$

A more explicit formula for this value can be obtained if we assume a constant average wage for all individuals:

$$S^{(i)}(t) = \bar{S}(t) = \text{uniform salary at time } t.$$

Taking into account the number of individuals in each age group (denoting by $L(x, t)$ the number of individuals of age x at time t in the population), we have

- For contributions collected at time t:

$$\pi \bar{S}(t) \sum_{x=x_0}^{x_r-1} L(x, t),$$

- For pension expenses at time t:

$$VP(t) \sum_{x=x_r}^{\omega} L(x, t) NP(x_r, t - x + x_r),$$

where $NP(x_r, t - x + x_r)$ is the sum of the points accumulated at retirement age x_r for the generation retiring at time $t - x + x_r$.

So we obtain

$$VP(t) = \frac{\pi \bar{S}(t) \sum_{x=x_0}^{x_r-1} L(x, t)}{\sum_{x=x_r}^{\omega} L(x, t) NP(x_r, t - x + x_r)}.$$

8.1.3 Actuarial Equilibrium Generation by Generation

Since the value of the point is the same for everyone in the technique developed above, and the number of points is fixed at retirement age, the result is that the annuity payments actually received by beneficiaries after they retire will not necessarily change in line with a natural revaluation rate on a price or salary index. Depending on demographic changes, pensions may even decrease over time in nominal term. In order to alleviate this problem and to protect the purchasing power of pensioners, a predetermined form of revaluation can be imposed after retirement.

In this case, the value of the point only concerns the generation retiring in the year. There is therefore a separate calculation per generation.

Total plan expenditures then become

$$VP(x_r,t) L(x_r,t) NP(x_r,t) + \sum_{x=x_r+1}^{\omega} VP(x,t)L(x,t)NP(x_r, t-x+x_r),$$

where $VP(x_r,t)$ is the value of the point for the generation retiring in year t and $VP(x,t)(x > x_r)$ is the value of the point for the generations already retired. This value is assumed to be indexed using a revaluation rate j_t and is therefore given recursively by

$$VP(x,t) = VP(x-1, t-1)(1+j_t).$$

Balancing revenues and expenses gives the following value of the point for the retiring generation:

$$VP(x_r,t) = \frac{\pi \bar{S}(t) \sum_{x=x_0}^{x_r-1} L(x,t) - \sum_{x=x_r+1}^{\omega} L(x,t) VP(x-1, t-1) \times (1+j_t) NP(x_r, t-x+x_r)}{L(x_r,t) NP(x_r,t)}.$$

This method puts all the demographic risk on the active members, since once they have retired, the members have a financial guarantee that their benefits will be indexed according to the chosen j index. The chosen revaluation factor is generally linked either to the evolution of a price index (the purchasing power of pensions is then maintained) or to the evolution of a wage index (the level of retirees then evolves in parallel with that of working people).

8.1.4 Contributions Call-Up Rate

Another technique for guaranteeing some stability in the benefits evolution is to introduce a decoupling between the contributions actually paid and the contributions taken into account in the calculation of points. In this case, there is in fact a mixture between DC and DB, since the contribution rate actually paid may fluctuate according to the general evolution of the scheme.

We have the following relationship

effective contribution rate = contractual contribution rate × contribution call rate,

8.1 Pay-As-You-Go DC Point Systems

where

- The effective contribution rate is the rate actually paid by the active population (either by social security contributions or by taxes, etc.).
- The contractual contribution rate is the rate taken into account in the allocation of retirement points.
- The contribution call rate is a corrective coefficient used to achieve a balance in the Pay-As-You-Go system, taking into account a constraint on the value of the point (for example, an automatic indexation rule). When this coefficient is less than 1, additional points can be granted without payment of contributions; on the contrary, when it is greater than 1, part of the contributions paid do not give any right to benefits.

Using a ξ call rate, the revenue becomes

$$\xi_t \pi \bar{S}(t) \sum_{x=x_0}^{x_r-1} L(x,t),$$

where π is the contractual contribution rate. If a single value of the point is determined (standard equilibrium), the rules for calculating the number of points are not changed; however, the value of the point becomes

$$VP(t) = \frac{\xi_t \pi \bar{S}(t) \sum_{x=x_0}^{x_r-1} L(x,t)}{\sum_{x=x_r}^{\omega} L(x,t) NP(x_r, t-x+x_r)}.$$

In a calculation by generation, we can also introduce this concept of call rate. From this point of view, we can consider three parameters that control the overall equilibrium of the plan: the value of the point for the new retirees of the year, the indexation rate of the old pensions and the call rate of the contributions. These three parameters (VP, j, ξ) obey the following actuarial equilibrium equation:

$$VP(x_r,t) = \frac{\xi_t \pi \bar{S}(t) \sum_{x=x_0}^{x_r-1} L(x,t) - \sum_{x=x_r+1}^{\omega} L(x,t) VP(x-1,t-1) \times (1+j_t) NP(x_r, t-x+x_r)}{L(x_r,t) NP(x_r,t)}.$$

Two remarks must be made. Firstly, given its virtual nature, the method makes it possible to grant additional points for periods prior to enrollment but recognized within the framework of the pension plan. This problem arises in particular when a plan is set up (the problem of the first generation); in this case, it is possible to reconstitute salaries and therefore points for both active and retired workers at the initial moment, even if the periods in question were not contributed. We find here again a strength of the Pay-As-You-Go system, which allows benefits to be granted at a consistent level, even in the start-up phase of the scheme.

On the other hand, rather than being a fixed percentage, the contribution can also be defined by slices, as in a traditional Defined Contribution plan.

8.1.5 Rate of Return of Point Systems

In order to measure the performance of a point system, we can introduce a notion of return. Here, the return is defined as the amount of pension earned in return for a unitary contractual contribution paid in the same year.

A unit contribution at time t will give right by definition to $\frac{1}{S_r(t)}$ points, and the corresponding pension will be given by

$$y(t) = VP(t)\frac{1}{S_r(t)}.$$

The yield is therefore the ratio between the value of the point and its acquisition price.

Assuming a uniform wage precisely equal to the reference wage, we can obtain the return, denoting by $D(t)$ the dependency ratio:

$$\begin{aligned}
y(t) &= \frac{VP(t)}{S_r(t)} \\
&= \frac{\pi \sum_{x=x_0}^{x_r-1} L(x,t)}{\sum_{x=x_r}^{\omega} L(x,t) NP(x_r, t-x+x_r)} \\
&= \frac{\pi \sum_{x=x_0}^{x_r-1} L(x,t)}{\sum_{x=x_r}^{\omega} L(x,t) \pi (x_r-x_0)} \\
&= \frac{\sum_{x=x_0}^{x_r-1} L(x,t)}{(x_r-x_0) \sum_{x=x_r}^{\omega} L(x,t)} \\
&= \frac{1}{(x_r-x_0) D(t)}.
\end{aligned}$$

The plan's performance is therefore purely demographic, inversely proportional to the dependency ratio.

In particular, for a stationary population, this return is constant. Table 8.1 gives the corresponding values of the return for various values of the dependency ratio and for a career duration of 40 years.

Table 8.1 Rate of return as a function of the dependency ratio

Dependency ratio	Rate of return
0.30	8.33%
0.40	6.25%
0.50	5.00%
0.60	4.17%

8.2 Hybrid Point Systems

The point system can also be considered in a philosophy, not of Defined Contribution, but in a hybrid system between Defined Benefit and Defined Contribution, as conceptually presented in all generalities in Chap. 7. We present here one possible architecture in this framework. We also introduce in this section the possibility of allowing for some flexibility in terms of retirement age (possible individual decision).

8.2.1 Basic Equations

In this general context of a point system, three actuarial relationships govern the dynamics of the plan. Two concern the level of benefits: the definition of the first pension for the year's new retirees and the adjustment of current pensions for old retirees. The third relationship expresses the balance of Pay-As-You-Go financing and involves the contribution rate.

8.2.1.1 Definition of First Pension (Current Year New Retirees)

The retirement pension, for the retiring generation, can be put in the form

$$\text{retirement pension} = \text{points account} \times \text{value of the point} \times \text{age coefficient} \tag{8.3}$$

or

$$R^{(i)}(t) = NP^{(i)} VP(t) ac^{(i)}(t),$$

where:

1. The points account $N^{(i)}$ is the sum of points acquired during the entire career: each year of an individual's career i entitles him/her to a number of points equal to

$$n^{(i)}(t) = \frac{S^{(i)}(t)}{\overline{S}(t)},$$

where $S^{(i)}(t)$ is the wage of individual i in year t (possibly capped) and $\overline{S}(t)$ is the average wage of the population in year t.

Note that since this is no longer a Defined Contribution scheme, it has been chosen to express the points on the basis of a salary and not a contribution paid, as in Sect. 8.1 (see Relation (8.1)).

The points account at the time of retirement will be given by

$$NP^{(i)} = \sum_{t=t_0(i)}^{t_1(i)-1} n^{(i)}(t),$$

where $t_0(i)$ is the affiliate's affiliation year i and $t_1(i)$ is the affiliate's retirement year.
2. The value of the point $VP(t)$ is the value of the conversion of points into euros for the generation retiring at time t; this value is identical for all members concerned. We will come back to the calculation of this value in more detail in Sect. 8.2.2.
3. The age coefficient $ac^{(i)}(t)$ is the coefficient allowing for the age at which individual i decides to retire. This coefficient ensures the (near) actuarial neutrality of the scheme. It is equal to one if the member leaves at his normal retirement age; it is less than one if he anticipates his retirement. This coefficient can, for example, be expressed as a ratio of life expectancy at two ages:

$$\text{age coefficient} = \frac{\text{life expectancy at normal retirement age}}{\text{life expectancy at actual retirement age}}.$$

This coefficient depends both on the member i (through his decision as to his effective retirement age) and on the time t of calculation (evolution over time of life expectancies). We will come back to the calculation of this coefficient in more detail in Sect. 8.2.3.

8.2.1.2 Current Pensions Adaptation (Current Year Old Retirees)

Pensions in payment are adjusted each year according to the following recurrent formula:

$$R^i(t+1) = R^i(t)(1 + g_{t+1})\beta_{t+1},$$

where g_{t+1} is the wage growth between t and $t+1$ such that

$$1 + g_{t+1} = \frac{\overline{S}(t+1)}{\overline{S}(t)},$$

and β_{t+1} is the sustainability coefficient applied at time $t+1$.

When the sustainability coefficient is equal to 1 ($\beta = 1$), pensions can be fully revaluated as wages.

8.2.1.3 Pay-As-You-Go Actuarial Equilibrium

There must be a balance between the contributions paid and the benefits paid:

$$\text{contribution rate} \times \text{payroll} = \text{sum of pensions paid}. \tag{8.4}$$

We will denote by π_t the contribution rate in year t.

The system is therefore based on 4 parameters: the value of the point, the age coefficient, the sustainability coefficient and the contribution rate.

We will explain in Sect. 8.2.4 how all these parameters can be managed dynamically while respecting the actuarial equilibrium relationship (8.4). First, let us detail how the value of the point and the age coefficient are calculated.

8.2.2 Calculation of the Value of the Point

The value of the point fixed in a given year only concerns the generation retiring in that year. In order to give a concrete meaning and ambition to this value, we can link it to the concept of benefit ratio. Consider a representative member i^* who has worked exactly at retirement age a prescribed number of reference years denoted N^* and who has earned each year a wage exactly equal to the average wage of the population; he will thus have acquired one point each year. At retirement he will have acquired N^* points. Since he has worked the required number of years, his age coefficient is equal to one. Taking into account (8.3), his first pension then becomes simply

$$R^{(i^*)}(t) = N^* \, VP(t).$$

Moreover, this pension can be expressed as a certain percentage of the average reference wage at that time; this corresponds for the representative member to his or her benefit ratio, denoted $\delta^*(t)$ (ratio between the first pension and the last wage). In the following, we will call this coefficient the target benefit ratio. We therefore also have:

$$R^{(i^*)}(t) = \delta^*(t)\overline{S}(t). \tag{8.5}$$

Equating these two quantities, we get for the value of the point:

$$VP(t) = \frac{\delta^*(t)\overline{S}(t)}{N^*}.$$

The value of the point can therefore be defined according to an ambition in terms of benefit ratio.

In particular, in a Defined Benefit plan, the benefit ratio is held constant over time ($\delta^*(t) = \delta^*$) and we have

$$VP(t) = \frac{\delta^* \overline{S}(t)}{N^*}. \tag{8.6}$$

The value of the point in this case evolves exactly at the same rate as the average salary.

In general, the target benefit ratio can be dynamic over time and follow different types of evolution depending on the chosen policy of parameter adaptation (see Sect. 8.2.4 below).

Given this value of the point, the first pension of a member leaving at his normal retirement age (age coefficient assumed to be one) will be according to (8.3):

$$R^{(i)}(t) = NP^{(i)} VP(t) = \frac{NP^{(i)}}{N^*} \delta^*(t) \overline{S}(t).$$

His pension differs from the average pension (8.5) because of his contributory power through the corrective ratio $NP(i)/N^*$.

8.2.3 Age Coefficient and Normal Retirement Age

The age coefficient ensures the actuarial neutrality of the system by comparing the normal age at which the member should have retired with the actual retirement age.

The normal retirement age of a member can be defined uniformly for all, as in traditional social security schemes (e.g., 67 years for all); such a strategy may, however, be anti-redistributive insofar as many studies have shown that life expectancy is very largely correlated with the social level of the member (differences in life expectancy at retirement can be significant). In order to correct this inequality, we could change the metric: instead of relying on age to verify whether a career is considered complete or not, we take into account the length of career. We then define a reference career length N^* that is identical for all. The normal retirement age then varies from one individual to another, depending on the age at which he or she started working.

In this framework, we denote by N^* the reference career length, by $x_0^{(i)}$ the career starting age for member i, by $x_r^{(i)} = x_0^{(i)} + N^*$ the normal retirement age for member i, and by $x_e^{(i)}$ the actual retirement age for member i.

One possible natural definition of the age coefficient is to express it as a ratio between two life expectancies (average number of years of survival remaining at a given age):

$$ac^{(i)}(t) = \frac{e\left(x_r^{(i)}, t\right)}{e\left(x_e^{(i)}, t\right)}, \tag{8.7}$$

8.2 Hybrid Point Systems

where

$$e(x,t) = \sum_{y=x+1}^{\infty} p_t(x,y),$$

is the life expectancy at age x at time t, with $p_t(x,y)$ being the probability that someone who was alive at age x, at time t, will still be alive at age y.

Note that Formula (8.7) implicitly takes into account three effects:

1. Social effect of longevity (parameter $x_r^{(i)}$): the correction is made not in relation to a uniform retirement age, but in relation to an age adapted to each individual according to his or her entry into working life.
2. Temporal effect of longevity (parameter t): these life expectancies change over time and are specific to each generation.
3. Individual accountability (parameter $x_e^{(i)}$): the member has flexibility in terms of retirement age (individual choice) but his pension is adjusted accordingly.

Taking into account the value of the point (8.6) and the conversion coefficient (8.7), the first pension (8.3) will be given by

$$R^{(i)}(t) = NP^{(i)}VP(t)ac^{(i)}(t) = \frac{NP^{(i)}}{N^*} \frac{e\left(x_r^i,t\right)}{e\left(x_e^i,t\right)} \left(\delta^*(t)\overline{S}(t)\right).$$

This pension can be expressed as the product of three factors:

1. The reference amount $\delta^*(t)\overline{S}(t)$, which is the target for the representative member (see (8.5)) (average retirement level expressed as a specified proportion of average earnings). This amount is representative of the desired level of generosity of the plan through the target benefit ratio $\delta^*(t)$.

    This collective term is then corrected to take into account the member's personal data concerning, on the one hand, his or her contributory capacity before retirement and, on the other hand, the duration during which the benefits are paid after retirement.
2. The ratio $\frac{NP^{(i)}}{N^*}$ expresses the contributory effort of the member throughout his career (comparison between the number of points acquired and the standard number corresponding to the length of the reference career); this value depends both on the level of the member's own earnings and the number of years worked.
3. The ratio $\frac{e(x_r^i,t)}{e(x_e^i,t)}$ adjusts benefits for the length of time they will be paid.

8.2.4 Automatic Adaptation of the Parameters

In order to maintain the actuarial equilibrium (8.4), we have three dynamic parameters to determine each year: the contribution rate $\pi(t)$, the target benefit ratio $\delta^*(t)$, the sustainability coefficient β_t.

We will assume here that the duration N^* is fixed. We will study in Sect. 8.2.5 the possible adaptation of this parameter. Given the number of variables involved, there are an infinite number of mechanisms for automatically adjusting the three parameters involved over time in order to ensure actuarial equilibrium at all times.

As an example, we can already highlight the following two extreme adjustments, corresponding respectively to a DB plan and a DC plan:

1. DB plan: the only adjustment variable is the contribution rate; the other variables are held constant:

$$\delta^*(t) = \delta^*,$$

$$\beta_t = 1.$$

2. DC plan: the contribution rate is kept constant; the benefit ratio is variable and the sustainability coefficient can be less than one:

$$\pi(t) = \pi^*,$$

$$\beta_t \leq 1.$$

There are, of course, an infinite number of other adaptation rules leading to different risk-sharings between generations (see Chap. 7) and that involve all three variables. As an example, let us consider the application of the Musgrave rule with proportional sharing of risk between generations (see Sect. 7.2.1). In this case, two additional rules are added to the actuarial equilibrium in order to univocally set the three parameters of the automatic adjustment.

Rule 1: Value of the Sustainability Coefficient

$$\beta_t = \frac{\delta^*(t)}{\delta^*(t-1)}. \tag{8.8}$$

This choice leads to a variation in current pensions at the same rate as the value of the point granted to new retirees. Indeed, we then have similar recursive relations for the two variables:

- For the value of the point:

$$VP(t) = \frac{\delta^*(t)\overline{S}(t)}{N^*} = \frac{\delta^*(t)}{\delta^*(t-1)} \frac{\overline{S}(t)}{\overline{S}(t-1)} VP(t-1).$$

- For the adjustment of current pensions:

$$R(t) = R(t-1)\frac{\overline{S}(t)}{\overline{S}(t-1)}\beta_t = \frac{\delta^*(t)}{\delta^*(t-1)} \frac{\overline{S}(t)}{\overline{S}(t-1)} R(t-1).$$

8.2 Hybrid Point Systems

Rule 2: Value of the Benefit Ratio

The target benefit ratio, which, given Rule 1 above, is the average benefit ratio for all retirees, follows the Musgrave rule:

$$\frac{\delta^*(t)}{1-\pi(t)} = \frac{\delta^*(t-1)}{1-\pi(t-1)}. \tag{8.9}$$

As for the actuarial balance, we have

$$\pi(t) = \delta^*(t)D(t). \tag{8.10}$$

Equations (8.8), (8.9) and (8.10) then allow us to adjust each year the 3 essential parameters of the plan (π, δ^* and β) according to the evolution of the dependency ratio.

8.2.5 Automatic Adjustment of the Reference Career Duration

Independently of the contribution rate, the target replacement rate or the sustainability rate, the reference career duration N^* can also become variable over time and constitute an annual adjustment parameter to take into account the increase in life expectancy. As an example, the following automatic adjustment rule can be provided for the reference duration N^*:

$$N^*_{t+1} = N^*_t \left(1 + \gamma_t \left(\frac{e(\tilde{x}, t)}{e(\tilde{x}, t-1)} - 1\right)\right),$$

where $e(\tilde{x}, t)$ is the life expectancy at a given age $\tilde{x}$ at time t and γ_t is a coefficient measuring the degree of adaptation.

When the coefficient γ is zero, there is no adaptation to life expectancy; when the coefficient is one, there is a full adaptation of career length to life expectancy. When the coefficient γ is strictly between 0 and 1, there is a partial adaptation.

It should be noted that in the points-based system described here, an adjustment of the reference duration has a double consequence. On the one hand, it delays the time when full retirement is granted through its influence on the normal retirement age and the age coefficient (8.7).

On the other hand, for members who still wish to retire at the age previously envisaged, an increase in the duration N^* will have a double effect on the amount of the pension: the age coefficient ac will be less than one and the value of the point will be reduced because of the influence of N^* on its value (see (8.6)).

Moreover, adapting the reference duration can also lead to modifying the automatic adaptation rules of the other parameters, such as for example the one proposed in Eq. (8.8). In the case of a joint adaptation of the reference duration

N^* and the target benefit ratio δ^*, we can, for example, adapt the sustainability coefficient β, by relating the benefit ratio to its reference duration, the recursive Eq. (8.6) then becoming

$$\beta_t = \frac{\delta^*(t)/N_t^*}{\delta^*(t-1)/N_{t-1}^*}.$$

This rule expresses that the sustainability coefficient evolves at the same rate as the target benefit ratio per reference year. As soon as one of the two parameters N^* or δ^* is adapted for the new retirees in a less advantageous direction (growth of N^* and/or decrease of δ^*), the sustainability coefficient becomes less than one. Other joint management rules for all these parameters can of course be envisaged, in a spirit of more or less pronounced risk-sharing between, on the one hand, the active and the retired and, on the other hand, the new retirees and the old retirees.

Part III
Fully Funded Pension Schemes

This third part is devoted to an in-depth analysis of fully funded pension schemes. These methods play a fundamental role in occupational pensions (second pillar). After recalling the logic of funding, we detail different funding methods, whether individual or collective. Actuarial techniques for amortizing gains and losses are also discussed.

Chapter 9
General Logic of Fully Funded Pension Plans

Abstract While there are many different fully funded methods, they share some common features. This chapter introduces these general concepts in the context of Defined Benefit schemes. The basic criteria for an initial classification of methods are introduced. Then the general logic of funded methods is discussed. The definitions presented in this chapter will be used systematically in Chaps. 10 and 11 devoted to the detailed description of the main individual or collective funded methods.

9.1 Characteristics and Classification of Funded Methods

Funded methods, in contrast to Pay-As-You-Go methods, aim to pre-finance the pension promise during the period of employment in such a way that, at retirement age, sufficient reserves exist to fully cover the pension commitment. This specific feature of these methods makes it particularly suitable for supplementary pension schemes set up by companies, for example. Indeed, and contrary to social security, the hypothesis of the scheme ending with the disappearance of the organizer cannot be ruled out in this case, and a funded method is the only way to protect the vested rights already accrued by retirees.

A funded method is therefore characterized by positive asset reserves (see Relationship (4.7)):

$$V(t) = V_r(t) + V_a(t),$$

where $V_r(t)$ is the aggregate reserve required to fully fund the benefits of those already retired and $V_a(t)$ is the balance of reserves, owned by active workers.

By definition, funded methods generate strictly positive workers' reserves, since they always have at least the totality of the retirees' reserves at their disposal. There is not one funded method, but rather an infinite number of methods. Two main criteria characterize the different methods: the speed of reserve evolution and the individual or collective nature of the method.

© The Author(s), under exclusive license to Springer Nature Switzerland AG 2025
P. Devolder, S. de Valeriola, *Actuarial Pension Funding Theory*, Springer Actuarial Textbooks, https://doi.org/10.1007/978-3-031-85268-8_9

9.1.1 The Speed of Reserve Evolution

During the period of employment, a funded method will lead to pre-financing the amount needed at retirement age to pay the benefits. The speed at which this pre-financing is formed during the active period can obviously vary considerably, as illustrated by Example 9.1.

> **Example 9.1** We make the following assumptions: Defined Benefit pension scheme, retirement age equal to 65 years, and age of affiliation equal to 25 years. We consider two methods. In the first one, called "initial finding", a single premium is paid at age 25 to fund the entire benefits. In the second one, called "late funding", the benefits are funded by periodic payments between the ages of 60 and 65.
>
> These two methods, both of which are funded methods since they generate strictly positive workers' reserves, are clearly contrasted and extreme. The maximum reserve formation speed of the first one can be opposed to a very slow pace for the second one.

It should be noted that this phenomenon of progressive and variable speed provisioning of a pension benefit can be compared to the accounting methods of amortization of an asset: different amortization methods are used and differ in their speed.

It could thus be considered that funded methods are simply the application of judicious amortization principles to the pension building, representing the amortization of a debt (the pension benefit).

9.1.2 Individual or Collective Nature

There are two main families of funded methods according to their degree of solidarity or individualization in the funding process.

In individual funded methods, the overall reserves of the scheme $V(t)$ can at any time be individualized per member:

$$V(t) = \sum_{j \in \mathcal{J}} V_j(t),$$

where $V_j(t)$ is the reserve of individual j and $\mathcal{J}$ is the set of all individuals.

In collective funded methods, global reserves cannot be individualized and serve the collective as a whole.

9.1 Characteristics and Classification of Funded Methods

Starting from the prospective relation of the global reserve (cf. relation (4.4)), we have:

$$V(t) = \sum_{s=t}^{\infty} SR(s)(1+i)^{-(s-t)} - \sum_{s=t}^{\infty} \pi(s) SS(s)(1+i)^{-(s-t)}.$$

Of course, in this collective relationship, both benefits and salaries can be individualized, so we can write:

$$V(t) = \sum_{s=t}^{\infty} \sum_{j \in \mathcal{J}} ER_j(s)(1+i)^{-(s-t)} - \sum_{s=t}^{\infty} \pi(s) \sum_{j \in \mathcal{J}} ES_j(s)(1+i)^{-(s-t)}, \quad (9.1)$$

where

- $ER_j(s)$ is the estimated value of the benefit paid to the individual j at time s, given by

$$ER_j(s) = \begin{cases} R_j(s) \cdot p_t(x_j, x_j + s - t) & \text{for } s \geq t + x_r - x_j, \\ 0 & \text{otherwise}, \end{cases}$$

where x_j is the age of individual j at time t and $R_j(s)$ is the benefit of individual j to be paid at time s,
- $ES_j(s)$ is the estimated value of individual j's salary at time s, given by

$$ES_j(s) = S_j(s) \cdot p_t(x_j, x_j + s - t) \quad \text{for} \quad t \leq s < t + x_r - x_j.$$

This global reserve can be broken down into a reserve for active workers and a reserve for retirees. In this case, the population at time t is decomposed into two sub-groups: $\mathcal{J}_r$ is made up of individuals who are retired at the time of computation, i.e. such that $x_j \geq x_r$; $\mathcal{J}_a$ is made up of individuals before retirement age, i.e. such that $x_j < x_r$. We then have

- For the retirees' reserve:

$$V_r(t) = \sum_{s=t}^{\infty} \sum_{j \in \mathcal{J}_r} ER_j(s)(1+i)^{-(s-t)} = \sum_{j \in \mathcal{J}_r} \sum_{s=t}^{\infty} ER_j(s)(1+i)^{-(s-t)}.$$

This reserve can therefore always be individualized by definition,
- For the workers' reserve:

$$V_a(t) = \sum_{s=t}^{\infty} \sum_{j \in \mathcal{J}_a} ER_j(s)(1+i)^{-(s-t)} - \sum_{s=t}^{\infty} \pi(s) \sum_{j \in \mathcal{J}_a} ES_j(s)(1+i)^{-(s-t)}.$$

This last relationship (and therefore de facto Eq. (9.1)) can only be individualized if the contribution rate π is also actuarially determined on an individual-by-individual basis and not for the whole group:

$$V_a(t) = \sum_{s=t}^{\infty} \sum_{j \in \mathcal{J}_a} ER_j(s)(1+i)^{-(s-t)} - \sum_{s=t}^{\infty} \sum_{j \in \mathcal{J}_a} \pi_j(s) ES_j(s)(1+i)^{-(s-t)}$$

$$= \sum_{j \in \mathcal{J}_a} \left[\sum_{s=1}^{\infty} ER_j(s)(1+i)^{-(s-t)} - \sum_{s=t}^{\infty} \pi_j(s) ES_j(s)(1+i)^{-(s-t)} \right]$$

$$= \sum_{j \in \mathcal{J}_a} V_j(t),$$

where $V_j(t)$ is the individual reserve of individual j (active worker).

The actuarial equilibrium is then imposed individual by individual: in this case we are in an individual funded method. Mathematically, the condition therefore comes down to being able to invert the temporal summation index and the summation index on the individuals in Eq. (9.1).

On the other hand, when the method defines the contribution rate not individually, but collectively, such a decomposition is meaningless and we then face a collective funded method: the retirees' reserve is always individualizable. However the workers' reserve can only be expressed globally.

9.2 Basic Quantities for Funded Methods

For all funded methods, either individual or collective, a certain number of basic quantities can be defined characterizing the funding process. Any funding method generates two mechanisms that need to be quantified: a mechanism for setting annual costs, i.e. the contributions to be paid, and a mechanism for the progressive formation of reserves, particularly workers' reserves. This will generate both an asset and a liability phenomenon.

We define below these basic quantities whose calculation methods, which vary according to the method considered, will be described in detail in Chaps. 10 and 11.

Accrued Liability

Global reserve representing the debt to members. It is therefore a balance sheet liability. $V(t) = AL(t)$ is the aggregate value at time t of the scheme's actuarial liability.

Assets/Fund

Market value of the financial assets backing the liabilities. It is therefore an asset balance sheet item. We will denote by $F(t)$ its value at time t.

9.2 Basic Quantities for Funded Methods

Normal Cost

Periodic contribution to be paid according to standard conditions generated by the chosen funding method. We will denote by $NC(t)$ its value at time t.

Unfunded Accrued Liability

The unfunded portion of the actuarial liability. It is at any time the difference between Accrued Liability and Assets. We will denote by $UAL(t)$ its value at time t.

Adjustment

Supplemental contribution over the normal cost to cover the unfunded liability. We will denote by $ADJ(t)$ its value at time t.

Global Cost

Total contributions payable resulting from the normal cost and adjustment contribution. We will denote by $C(t)$ its value at time t.

Let us detail these different actuarial elements.

"Accrued Liability" and "Normal Cost" are typically liabilities, defined on the basis of actuarial assumptions. These two concepts are linked by the prospective relationship of the reserves:

$$AL(t) = \sum_{s=t}\sum_{j} ER_j(s)(1+i)^{-(s-t)} - \sum_{s=t} NC(s)(1+i)^{-(s-t)} \qquad (9.2)$$

$$= PVFB(t) - PVFNC(T), \qquad (9.3)$$

where $PVFB(t)$ is the discounted value of future benefits, and $PVFNC(t)$ is the discounted value of future normal costs.

The assets held may differ from the actuarial liabilities for a variety of reasons. Depending on their cause, there are three types of differences:

- First-order differences: differences resulting from the gap between the actuarial assumptions used in the calculation of liabilities and the actual situation or from a change in assumptions (e.g. difference between the theoretical life table and actual mortality, difference between the actuarial discount rate and the actual rate of return on assets, etc.).
- Second-order differences: differences resulting from the lack of funding for recognized past service prior to the inception of the scheme.
- Third-order differences: other types of differences that may arise from the funding method.

So the Unfunded Accrued Liability (that can be positive or negative) is given by

$$UAL(t) = AL(t) - F(t).$$

As for the adjustment contribution, it can be either implicit or explicit. It is implicit when it is included in the method's calculation mode itself. It is explicit when it has to be highlighted. It can then be the subject of a one-time amortization or an amortization spread over time:

- one-time adjustment: $ADJ(t) = UAL(t)$,
- adjustment spread over time, for example: $ADJ(t) = \frac{UAL(t)}{\ddot{a}}$, where $\ddot{a}$ is a financial or life annuity over a fixed period of time.

The methods for amortizing actuarial gains and losses will be detailed in Chap. 12.

The total cost of the pension plan is of course given by

$$C(t) = NC(t) + ADJ(t).$$

In an individual funded method, the principle of actuarial equilibrium is applied on an individual-by-individual basis and all these basic quantities are individualized by individual. One can write for example:

$$AL(t) = \sum_{j \in \mathcal{J}} AL_j(t), \qquad NC(t) = \sum_{j \in \mathcal{J}} NC_j.$$

In a collective funded method, these basic quantities remain essentially collective and are not subject to any individualization. Only equilibrium at the collective level is sought.

Chapter 10
Individual Funding Methods

Abstract In this chapter we describe the different individual funding techniques used in the financing of Defined Benefit pension schemes. For each of the methods presented, we explain how the contributions to be paid and the reserves to be constituted are calculated. We also illustrate the funding logic by addressing the concepts of actuarial anticipation and pension gap corresponding to early retirement and mobility during the career.

10.1 General Principles of Individual Funded Methods

We will consider here a fully funded Defined Benefit pension scheme, as defined in Chap. 9. Each individual therefore funds his own retirement on his own account. All the traditionally collective concepts in a pension scheme, such as reserves or contributions, are individualized by member.

The basic principle of individual funding is that each individual must have fully funded the pension liability to be paid from the age of retirement during his working life. There are many individual funding methods, each characterized by its own pace of building up the necessary reserves and therefore a different distribution of funding over time.

Three criteria can be used in order to map the main families of fully funded methods:

- First criterion: funding with or without projection,
- Second criterion: funding by successive single premiums or by level premiums,
- Third criterion: whether or not past services are funded separately.

10.1.1 First Criterion: Funding with or Without Projection

Methods without projection only take into account financial data known at the time of calculation. In particular, the amount of salaries as known at the date of calculation is used.

Methods with projections take into account projections of financial data at retirement age. In particular, future projected salaries at retirement age are used in the computation today.

10.1.2 Second Criterion: Funding by Successive Single Premiums or by Constant Premiums

The successive single premium methods (unit credit cost methods), aim to divide the pension commitment into equal units to be funded each year in single premiums (full annual funding of fixed portions of the benefits). The constraint therefore relates to the actuarial liability, which is in a way the independent variable. The normal cost will result as the dependent variable. These methods are also sometimes referred to as "liability-driven methods" since they aim to spread the constitution of liabilities evenly throughout the career.

Constant premium methods (level premium methods) aim to fund the commitment by constant or stable annual payments. The constraint is therefore this time on the funding, and the actuarial liability will result. These methods are also sometimes referred to as "cost allocation methods" or "cost driven methods" since they aim to spread the funding evenly throughout the career.

This difference in approach can already be illustrated simply by looking at a pure endowment transaction that one wishes to fund by periodic payments. The traditional solution used in life insurance is to work with constant premiums. By considering a unit capital payable in case of life at age $x + n$ (pure endowment, see technical appendix), the constant premium p to be paid in advance each year between ages x and $x + n$ is given by

$$p = \frac{{}_nE_x}{\ddot{a}_{x,\overline{n}|}}.$$

This method, interpreted in terms of liability, leads to funding a portion of the capital each year. The premium paid in the year t $(0 \leq t < n)$ makes it possible to fund a portion of capital denoted K_t and given by

$$K_t = \frac{p}{{}_{n-t}E_{x+t}} = \frac{{}_nE_x}{{}_{n-t}E_{x+t}\,\ddot{a}_{x,\overline{n}|}}.$$

10.1 General Principles of Individual Funded Methods

In particular the first element is given by

$$K_0 = \frac{1}{\ddot{a}_{x,\overline{n}|}} > \frac{1}{n}.$$

Given the growth in the price of the pure endowment, the K function is decreasing. The constant premium method therefore generates a decreasing amortization process. This alternative interpretation naturally leads to consider another form of funding: rather than considering constant premiums, one wishes to obtain a linear amortization of the capital.

Each year the p_t premium to be paid will be the solution to the equation

$$K_t = \frac{p_t}{{}_{n-t}E_{x+t}} = \frac{1}{n}.$$

This time the premium is therefore variable, increasing over time and given by

$$p_t = \frac{1}{n} {}_{n-t}E_{x+t}.$$

This alternative approach to funding a pure endowment transaction in periodic premiums corresponds to the method of successive single premiums in pension theory.

10.1.3 Third Criterion: Whether or Not Past Services are Funded Separately

When the plan recognizes prior services (e.g. years counted since entry into service) on joining the plan, some methods separately fund this initial past service using another funding method. This is referred to as explicit amortization of the initial back service. In other methods, initial past and future services are globally considered.

10.1.4 Crossing the Criteria

By crossing these three criteria we can generate the main individual funded methods (Table 10.1).

In order to compare the different methods, we will apply them to a standard pension scheme, which is designed to provide a pension for a 40-year career equal to 50% of the last working salary, payable from the age of 60.

Table 10.1 Individual funded methods

	Unprojected	Projected
Successive single premiums	Unit Credit Cost (UC)	Projected Unit Credit Cost (PUC)
Constant premiums	Individual Level Premium (ILP)	Projected Individual Level Premium (PILP)
		Projected Individual Level Percent (PLP)
Constant premiums with explicit back service amortization	Individual With Supplemental Liabilities (ISL)	Projected Individual Supplemental Liabilities (PISL)
	Normal Entry Age (EA)	Projected Normal Entry Age (PEA)

For each member of the scheme, at the age of 60 a reserve given by

$$V_{60} = \frac{N}{40} 50\% S(60) \cdot \ddot{a}_{60}$$

must be fully funded, where N is the career length taken into account ($N \leq 40$), $S(60)$ is the last working wage and

$$\ddot{a}_{60} = \text{annuity price at 60 years} = \sum_{x=60}^{\infty} \frac{\ell_x}{\ell_{60}} \frac{(1+j)^{x-60}}{(1+i)^{x-60}},$$

where ℓ_x is the mortality table, i is the discount rate, and j is the annuity revaluation rate.

Note that this annuity price coefficient will later be a simple multiplicative factor. One can easily consider other annuity prices, for example a monthly annuity price ($\ddot{a}_{60}^{(12)}$) or joint life.

10.2 Unit Credit Cost

10.2.1 General Principle

An affiliate who has started working and is affiliated at $t = 0$ at age x_0 is considered. The following diagram is summarized on the time scale in Fig. 10.1, where x_0 is the age at affiliation, $x = x_0 + t$ is the age at calculation time, x_r is the retirement age and $T = x_r - x_0$ is the retirement date.

The objective of the unit credit method is, taking into account the data observed at any given moment of calculation, to hold the reserve that will enable the benefits corresponding to the years already worked to be paid in full at retirement. In the

10.2 Unit Credit Cost

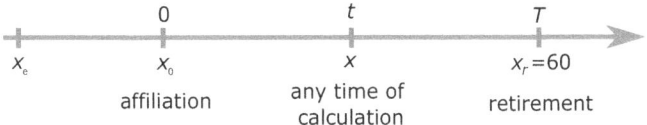

Fig. 10.1 Time line

following, we will assume the contributions payable annually in advance. The first contribution will therefore be paid at $t = 0$ and will cover the first year of service between $t = 0$ and $t = 1$. The last contribution will be paid at $t = T - 1$.

The funding philosophy is underpinned by the idea that each year worked entitles the holder to a complementary fraction of the pension obligation. The method therefore explicitly dictates the level of actuarial liabilities to be funded. Without back service, the reserve to be made at time t is given by

$$AL(t) = \frac{t}{40} 50\% S(t-1) \cdot \ddot{a}_{60} \cdot {}_{60-x}E_x \quad (0 < t \leq T). \tag{10.1}$$

The normal cost is the amount required to be paid at the beginning of the year t (prepayment assumption) to move the reserve amount from the $AL(t)$ level to the $AL(t+1)$ level at time $t+1$. It is therefore expressed as the difference between the discounted value of the actuarial liability to be accrued at time $t+1$ and the liability already accrued:

NC_t = normal cost to be paid at time t

$= AL(t+1) {}_1E_x - AL(t)$

$= \dfrac{t+1}{40} 50\% S(t) \ddot{a}_{60} \cdot {}_{60-x-1}E_{x+1} \cdot {}_1E_x - \dfrac{t}{40} 50\% S(t-1) \ddot{a}_{60} \cdot {}_{60-x}E_x$

$$= \left[\underbrace{\frac{1}{40} 50\% S(t)}_{\text{cost of year}} + \underbrace{\frac{t}{40} 50\% (S(t) - S(t-1))}_{\text{valuing the past}} \right] \ddot{a}_{60} \cdot {}_{60-x}E_x. \tag{10.2}$$

In particular the first year we get

$$NC_0 = \frac{1}{40} 50\% S(0) \cdot \ddot{a}_{60} \cdot {}_{60-x_0}E_{x_0}.$$

The amount (10.2) will therefore increase over the course of the career for two reasons: growth in the price of the pure endowment (term ${}_{60-x}E_x$) and revaluation of the past with linear growth (term in $t/40$). The method thus generates a significant increase in costs during the career. Observe that even if the salary remains constant (no term of revaluation of the past in this case), the contributions will still increase due to the effect of the increase in the price of the pure endowment. It should be noted that the past adjustment term is linked to the type of scheme considered here (benefit on final salary). It disappears for fixed lump-sum schemes or non-indexed career average schemes.

10.2.2 Introducing a Back Service

The initialization of this method when enrolling in the plan is substantially modified when there is recognition of pre-affiliation service (back service).

On the one hand, if, as assumed above, there is no recognition of years prior to affiliation, the total length of the career is given by $N = T = x_r - x_0$. The first normal cost is then simply the funding of one year of service and is given by

$$NC_0 = \frac{1}{40} 50\% S(0) \cdot \ddot{a}_{60 \,\cdot\, 60-x_0} E_{x_0}.$$

On the other hand, if there is recognition of years of service prior to affiliation, the method will induce significant initial funding corresponding to the full catch-up of those years. This initial funding is called initial back service. Denoting by N_1 the number of years recognized prior to enrollment (e.g. $N_1 = x_0 - x_e$, where x_e is the age of entry into service in the company) the total length of the career is given by

$$N = N_1 + T = (x_0 - x_e) + (x_r - x_0).$$

The actuarial liability of the method at time t becomes:

$$AL(t) = \frac{N_1 + t}{40} 50\% S(t-1) \cdot \ddot{a}_{60} \cdot {}_{60-x} E_x.$$

In this case, the first normal cost, in addition to the funding of the current year, must also fully cover the burden of the past:

$$NC_0 = \frac{N_1 + 1}{40} 50\% S(0) \cdot \ddot{a}_{60} \cdot {}_{60-x_0} E_{x_0}.$$

In the same way the subsequent normal costs become

$$NC_t = \left(\frac{1}{40} 50\% S(t) + \frac{t}{40} 50\% (S(t) - S(t-1)) \right.$$
$$\left. + \frac{N_1}{40} 50\% (S(t) - S(t-1)) \right) \ddot{a}_{60} \cdot {}_{60-x} E_x.$$

These contributions are made up of three components: the cost of the year ($1/40$), the revaluation of recognized service since affiliation ($t/40$), the revaluation of recognized service prior to affiliation ($N_1/40$).

In this method, the recognition of pre-affiliation service therefore has spill-over effects in funding throughout the career and not just at affiliation.

10.3 Projected Unit Credit Cost

The principle is identical to that of the Unit Credit method, but in order to avoid as much as possible the effect of the revaluation term, we work with projected salaries.

At instant of calculation t, the salary $S(t)$ is replaced by a final wage estimator, for example

$$S^*(t) = S(t) \cdot (1+g)^{60-x-1},$$

where g is an estimator of the rate of salary growth.

Other more complex models for estimating future salaries can of course be implemented, taking into account, for example, pre-existing pay scales.

For example, with j = inflation rate, k = productivity rate (salary growth above inflation), $s(y)$ = salary scale at age y, we will have an estimator at time t of the final salary given by

$$S^*(t) = S(t) \cdot (1+j+k)^{60-x-1} \frac{s(x_r - 1)}{s(x)}.$$

Assuming an overall geometric progression with common ratio g, the variation in this projected salary from one year to the next is then given by

$$S^*(t) = S(t) \cdot (1+g)^{60-x-1}$$

$$S^*(t+1) = S(t+1) \cdot (1+g)^{60-x-2}$$

$$= S(t) \cdot \frac{S(t+1)}{S(t)} \cdot (1+g)^{60-x-2}$$

$$= S(t) \cdot (1 + g_r(t+1)) \cdot (1+g)^{60-x-2},$$

where $g_r(t+1)$ is the real salary growth rate between t and $t+1$.

Finally,

$$S^*(t+1) = S^*(t) \cdot \frac{1 + g_r(t+1)}{1+g}.$$

The two successive projected estimators coincide if the actual growth rate matches the projection rate.

The objective of the method is then to have at all times an actuarial liability allowing for the full payment at retirement of the benefits corresponding to the years already worked on the basis of a projected salary:

$$AL(t) = \frac{t}{40} 50\% S^*(t-1) \cdot \ddot{a}_{60} \cdot {}_{60-x}E_x \qquad (0 \leq t \leq T).$$

In terms of normal cost and without recognition of back service, we obtain

$$NC_t = \left[\frac{1}{40} 50\% S^*(t) + \frac{t}{40} 50\% \left(S^*(t) - S^*(t-1) \right) \right] \ddot{a}_{60} \cdot {}_{60-x}E_x$$

$$= \left[\frac{1}{40} 50\% S^*(t) + \frac{t}{40} 50\% S^*(t-1) \cdot \frac{g_r(t) - g}{(1+g)} \right] \ddot{a}_{60} \cdot {}_{60-x}E_x.$$

The second revaluation term is zero when reality follows the assumptions.

Similarly, when there is recognition of initial past service, the general normal cost formula becomes

$$NC_t = \left[\frac{1}{40} 50\% S^*(t) + \frac{N_1 + t}{40} 50\% S^*(t-1) \cdot \frac{g_r(t) - g}{(1+g)} \right] \ddot{a}_{60} \cdot {}_{60-x}E_x.$$

10.4 Individual Level Premium

Even with a static salary assumption, we have seen that the unit credit method leads to an increase in costs during the career. The objective of the individual level premium method is precisely to determine a contribution which, all other things being equal, will remain constant until retirement age. The focus is no longer on actuarial liabilities as in unit credit but on financing and its stability.

The first contribution, assumed to be payable annually in advance, will be given by the principle of actuarial equivalence:

discounted value of constant contributions during the career

$$=$$

discounted value of liabilities.

Hence

$$NC_0 \cdot \ddot{a}_{x_0, \overline{60-x_0|}} = \frac{N}{40} 50\% S(0) \ddot{a}_{60} {}_{60-x_0}E_{x_0}$$

$$\Rightarrow NC_0 = \frac{\frac{N}{40} 50\% S(0) \ddot{a}_{60} {}_{60-x_0}E_{x_0}}{\ddot{a}_{x_0, \overline{60-x_0|}}}. \qquad (10.3)$$

10.4 Individual Level Premium

We get the price of a pure endowment with constant annual premiums for a capital corresponding to the pension obligation :

$$NC_0 = \left(\frac{N}{40} 50\% S(0) \ddot{a}_{60}\right) \frac{_{60-x_0}E_{x_0}}{\ddot{a}_{x_0,\overline{60-x_0|}}}.$$

This premium will remain constant as long as the insured capital remains constant, i.e. as long as the salary remains constant.

In practice, the annual salary increases will require funding additional capital and therefore additional premiums to be paid. Thus, if at $t = 1$, the salary increases from $S(0)$ to $S(1)$, we will have

$$NC_1 = NC_0 + \Delta NC_1,$$

where ΔNC_1 is the constant annual premium to be paid from age $x_0 + 1$ to insure a capital equal to

$$\frac{N}{40} 50\% (S(1) - S(0)) \ddot{a}_{60}.$$

We have therefore

$$\Delta NC_1 = \left[\frac{N}{40} 50\% \Delta S(1) \ddot{a}_{60}\right] \frac{_{60-x_0-1}E_{x_0+1}}{\ddot{a}_{x_0+1,\overline{60-x_0-1|}}}.$$

These premium supplements are becoming more and more expensive due to a twofold effect: discounting over a shorter period and spreading over a smaller number of payments. Thus, what was supposed to remain constant ex ante no longer remains constant ex post because of the salary evolution.

The method has explicitly defined the level of normal costs. The actuarial liability is thus expressed, in accordance with the general relationship (9.2), as the difference between the discounted value of future benefits and the discounted value of future normal costs assumed by definition under the method to be equal to the last calculated normal cost (expressed at time t but before recalculation of the new normal cost at t):

$$AL(t) = \frac{N}{40} 50\% S(t-1) \ddot{a}_{60} \cdot {_{60-x_0-t}E_{x_0+t}} - NC_{t-1}$$
$$\cdot \ddot{a}_{x_0+t,\overline{60-x_0-t|}} \quad (0 < t \leq T). \tag{10.4}$$

Finally, it should be noted that this method can be applied irrespective of whether or not there is recognition of initial past service. Indeed, by definition, the entire commitment is taken into account from the outset, whether it concerns past or future services. However, when past services are to be funded separately, the "individual with supplemental liability" method described below is used.

10.5 Individual Level with Supplemental Liabilities

The "individual level with supplemental liabilities" method consists of breaking down the total funding of the scheme into two parts: the portion of the obligations relating to initial past services is funded in unit credit cost, while the balance of the obligations is funded in individual level premium.

In concrete terms, the initial commitments relating to services before affiliation are calculated as

$$K_1 = \frac{N_1}{40} 50\% S(0) \cdot \ddot{a}_{60}.$$

The normal cost of the method each year is then that obtained in individual level premium, after subtracting the fixed amount K_1 from the total commitments:

$$NC_0 = \left[\frac{N}{40} 50\% S(0) \cdot \ddot{a}_{60} - K_1 \right] \frac{_{60-x_0}E_{x_0}}{\ddot{a}_{x_0,\overline{60-x_0}|}}. \quad (10.5)$$

Subsequent normal costs are calculated using the same recurrence formulas:

$$NC_{t+1} = NC_t + \Delta NC_{t+1},$$

with

$$\Delta NC_{t+1} = \left[\frac{N}{40} 50\% \Delta S(t+1) \ddot{a}_{60} \right] \frac{_{60-x_0-t-1}E_{x_0+t+1}}{\ddot{a}_{x_0+t+1,\overline{60-x_0-t-1}|}}. \quad (10.6)$$

On the other hand, the amount K_1 must of course be the subject of a complementary funding given in unit credit by the following initial single premium which will be added the first year to the normal cost NC_0:

$$ADJ(0) = \frac{N_1}{40} 50\% S(0) \cdot \ddot{a}_{60} \cdot {_{60-x_0}E_{x_0}},$$

$$C(0) = NC_0 + ADJ(0).$$

In terms of actuarial liabilities, the general formula (10.4) for individual level premium still applies. Only the value of the normal cost differs (see relationships (10.5) and (10.6)). The normal cost will be lower under this method because of the full amortization of past service and the actuarial liability will therefore be higher.

It should also be noted that the initial adjustment cost $ADJ(0)$ can be spread over several years (for example, over a fixed 5-year term). When this cost is spread over the entire career, the basic individual level premium method is of course recovered.

Note that this method can be considered as a mix between the unit credit method for the initial frozen past service portion and the individual level premium method

10.5 Individual Level with Supplemental Liabilities

for the balance. A variant of the unit credit method described in Sect. 10.2 could be symmetrically considered, this time applying the individual level premium method for the initial frozen past service and the unit credit method for the balance. The objective is then to spread the cost of the initial past service that would normally have to be funded initially in the unit credit logic.

This method could be called "unit credit with supplemental liabilities". In this case, we again have an adjustment contribution. Explicitly we have, denoting by N_1 the initial back service:

- Normal cost :
  - first year:

$$NC_0 = \frac{1}{40} 50\% S(0) \cdot \ddot{a}_{60} \cdot {}_{60-x_0}E_{x_0},$$

  - subsequent years ($t > 0$):

$$NC_t = \left[\frac{1}{40} 50\% S(t) + \frac{N_1 + t}{40} 50\% S(t-1) \cdot \frac{g_r(t) - g}{(1+g)} \right] \ddot{a}_{60} \cdot {}_{60-x}E_x,$$

- Unfunded frozen initial liabilities:

$$UAL(0) = \frac{N_1}{40} 50\% S(0) \ddot{a}_{60} \cdot {}_{60-x_0}E_{x_0},$$

- Annual adjustment contribution: to be paid at $t = 0, 1, \ldots, T-1$:

$$ADJ(0) = \frac{UAL(0)}{\ddot{a}_{x_0, \overline{60-x_0|}}},$$

- Total contribution : for $t = 0, 1, \ldots, T-1$,

$$C(t) = NC_t + ADJ(0),$$

- Unfunded liabilities:

$$UAL(t) = ADJ(0) \ddot{a}_{x_0+t, \overline{60-x_0-t|}} = UAL(0) \frac{\ddot{a}_{x_0+t, \overline{60-x_0-t|}}}{\ddot{a}_{x_0, \overline{60-x_0|}}},$$

- Total actuarial liabilities (including unfunded liabilities) as in Unit Credit:

$$AL(t) = \frac{N_1 + t}{40} 50\% S(t-1) \cdot \ddot{a}_{60} \cdot {}_{60-x}E_x.$$

- The actuarial liabilities excluding the part unfunded is of course given by $AL(t) - UAL(t)$.

10.6 Projected Individual Level Premium

The principle is identical to that of the individual level premium method, but in order to avoid contribution increases in the course of a career, we work with projected salaries.

As in the projected unit credit method, the salary S is replaced by a projected salary S^*, given for example by

$$S^*(t) = S(t) \cdot (1+g)^{60-x-1}.$$

The first normal cost is given by:

$$NC_0 = \left[\frac{N}{40} 50\% S^*(0) \cdot \ddot{a}_{60}\right] \frac{_{60-x_0}E_{x_0}}{\ddot{a}_{x_0,\overline{60-x_0}|}}.$$

If real growth follows the assumption, the premiums obtained will this time be constant. Otherwise, it will be necessary to fund an additional capital given by:

$$\Delta NC_{t+1} = \left[\frac{N}{40} 50\% \left(S^*(t+1) - S^*(t)\right) \ddot{a}_{60}\right] \cdot \frac{_{60-x_0-t-1}E_{x_0+t+1}}{\ddot{a}_{x_0+t+1,\overline{60-x_0-t-1}|}}.$$

It should be noted that the method generally leads to particularly high contributions in the early years because of the anticipation of future inflation over the entire career (very significant pre-funding). In particular, the contribution rate (ratio between normal cost and salary) will decrease over time.

The actuarial liability is given by the following formula which directly generalizes the individual level premium formula by simply substituting projected elements in both the benefits and contributions parts:

$$AL(t) = \frac{N}{40} 50\% S^*(t-1) \ddot{a}_{60} \cdot {_{60-x_0-t}E_{x_0+t}} - NC_{t-1} \ddot{a}_{x_0+t,\overline{60-x_0-t}|}.$$

Note that under this method, initial past service can also be funded separately. The method is identical to that of the individual with supplemental liability, except that all elements are projected. A projected unit credit with supplemental liabilities method can also be considered.

10.7 Projected Individual Level Percent

The objective is similar to that of the projected individual level premium method, but this time stability of funding is sought not in absolute terms but as a percentage of salaries. Indeed, the projected individual level premium method leads to significant pre-funding and in particular to decreasing contribution rates.

10.7 Projected Individual Level Percent

This new method therefore aims to define a contribution rate on salary that is as stable as possible throughout the career and on an individual-by-individual basis. It is therefore a projected method. It should be noted that this method is known in its collective version as "aggregate cost" (see Sect. 11.3). The contribution rate will be obtained by the actuarial equivalence relationship

$$\text{discounted value of contributions at a constant rate}$$
$$=$$
$$\text{discounted value of projected liabilities.}$$

Hence

$$\pi \sum_{n=0}^{60-x_0-1} S(0) \cdot (1+g)^n \cdot \frac{1}{(1+i)^n} {}_n p_{x_0} = \frac{N}{40} 50\% S(0)(1+g)^{60-x_0-1} \cdot \ddot{a}_{60} \cdot {}_{60-x_0}E_{x_0}.$$

The contribution rate is given by:

$$\pi = \frac{\frac{N}{40} 50\%(1+g)^{60-x_0-1} \cdot \ddot{a}_{60} \cdot {}_{60-x_0}E_{x_0}}{\sum_{n=0}^{60-x_0-1}(1+g)^n \frac{1}{(1+i)^n} \cdot {}_n p_{x_0}}. \tag{10.7}$$

The first normal cost is given by

$$NC_0 = \pi \cdot S(0)$$
$$= \frac{\frac{N}{40} 50\% S(0)(1+g)^{60-x_0-1} \cdot \ddot{a}_{60} \cdot {}_{60-x_0}E_{x_0}}{\sum_{n=0}^{60-x_0-1}(1+g)^n \frac{1}{(1+i)^n} \cdot {}_n p_{x_0}}.$$

Let $\frac{1+g}{1+i} = \frac{1}{1+k}$, i.e. k = real interest rate. So we have

$$NC_0 = \frac{\frac{N}{40} 50\% S(0) \cdot \ddot{a}_{60} \cdot {}_{60-x_0}E_{x_0}^{(k)}}{\ddot{a}_{x_0,\overline{60-x_0}|}^{(k)}(1+g)},$$

where the actuarial elements $E^{(k)}$ and $\ddot{a}^{(k)}$ are computed at the actual interest rate k.

This formula is therefore similar (if we neglect the factor $(1+g)$) to the normal cost factor in individual level premium (cf. Relation (10.3)) except for the use of a real rate in the discount factors.

In subsequent years, if the reality corresponds to the assumptions, in particular with regard to salary growth, the rate will remain stable by definition and the normal cost will be given by

$$NC_t = \pi S(t) = \pi S(0)(1+g)^{x-x_0}.$$

The normal cost is therefore increasing in line with salaries.

The actuarial liability is given by

$$AL(t) = \frac{N}{40} 50\% S^*(t-1) \ddot{a}_{60} \cdot {}_{60-x_0-t}E_{x_0+t} - \pi S(0) \sum_{y=x}^{59} (1+g)^{y-x_0} {}_{y-x}E_x.$$

If real salary growth differs from the projection assumption, a new contribution rate must be calculated. The same recursive methodology can be used for this as for individual level premium: a change in the contribution rate is associated with a change in the projected salary:

$$\pi_{t+1} = \pi_t + \Delta \pi_{t+1},$$

where

$$\Delta \pi_{t+1} = \frac{\frac{N}{40} 50\% \left(S^*(t+1) - S^*(t)\right) \cdot \ddot{a}_{60} \cdot {}_{60-x_0-1}E_{x_0+1}}{S(t+1) \sum_{n=0}^{60-x-2} (1+g)^n \cdot {}_n E_{x+1}}.$$

10.8 Normal Entry Age

The individual level premium method generates additional contributions due to successive salary increases. These supplements are increasingly expensive due to the growth in the premium rate (cf. Relation (10.4)).

The idea of the normal entry age method is to disregard this aging process and to continue obstinately applying the initial price of the pure endowment at entry age in the calculation of successive normal costs. One could therefore say that an actuarial error is at the basis of the method. This error is of course not forgotten and leads to the identification of a deficit that will have to be amortized by an adjustment contribution in addition to the normal cost.

It is therefore once again a variant of the individual level premium method that can be used with or without projection. The method without projection will be developed further below.

Here we will generally assume that the entry age is different from the age of affiliation, which will cause a deficit from the start of funding, as can be seen in Fig. 10.2. The first contribution is given by

$$NC_0^{(1)} = C_0 \cdot \frac{{}_{60-x}E_x}{\ddot{a}_{x,\overline{60-x|}}},$$

10.8 Normal Entry Age

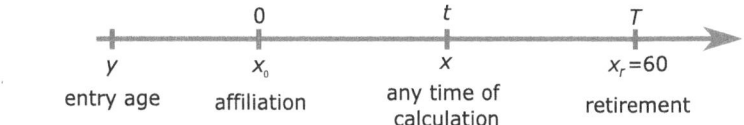

Fig. 10.2 Time line

where

$$C_0 = \frac{N_0}{40} 50\% S(0) \cdot \ddot{a}_{60}$$

in classic individual level, and

$$NC_0^{(2)} = C_0 \cdot \frac{_{60-y}E_y}{\ddot{a}_{y,\overline{60-y|}}}$$

in normal entry age. Of course we have $NC_0^{(2)} < NC_0^{(1)}$.

The payment of this constant amount $NC_0^{(2)}$ between the affiliation age and the retirement age thus causes an initial deficit given by

$UAL(0) =$ initial unfunded obligations

$=$ current value of liabilities - discounted value of future payments

$$= C_0 \cdot {_{60-x}E_x} - NC_0^{(2)} \cdot \ddot{a}_{x,\overline{60-x|}}$$

$$= C_0 \cdot {_{60-x}E_x} - C_0 \cdot \frac{_{60-y}E_y}{\ddot{a}_{y,\overline{60-y|}}} \cdot \ddot{a}_{x,\overline{60-x|}}$$

$$= C_0 \left({_{60-x}E_x} - {_{60-y}E_y} \cdot \frac{\ddot{a}_{x,\overline{60-x|}}}{\ddot{a}_{y,\overline{60-y|}}} \right)$$

$$= \frac{C_0}{\ddot{a}_{y,\overline{60-y|}}} {_{60-y}E_y} \left(\frac{_{60-x}E_x}{_{60-y}E_y} \ddot{a}_{y,\overline{60-y|}} - \ddot{a}_{x,\overline{60-x|}} \right)$$

$$= NC_0^{(2)} \left(\frac{1}{_{x-y}E_y} \ddot{a}_{y,\overline{60-y|}} - \ddot{a}_{x,\overline{60-x|}} \right)$$

$$= \frac{NC_0^{(2)}}{_{x-y}E_y} \ddot{a}_{y,\overline{x-y|}}.$$

This amount is equal to the reserve that would have been accumulated at age x if there had been constant premium payments $NC_0^{(2)}$ since entry age y. This deficit

$UAL(0)$ is amortized either in one lump sum or in equal installments (e.g. over 5 years):

$$ADJ(0) = UAL(0)$$

or

$$ADJ(0) = \frac{UAL(0)}{\ddot{a}_{x_0,\overline{m}|}}.$$

The total contribution in normal entry age is

$$C(0) = NC_0^{(2)} + ADJ(0).$$

In the following years, wage increases again cause deficits that have to be amortized. At time $t+1$, the normal cost is

$$NC_{t+1} = \left(\frac{N}{40} 50\% S(t+1)\ddot{a}_{60}\right) \cdot \frac{_yE_{60-y}}{\ddot{a}_{y,\overline{60-y}|}}.$$

The price of the pure endowment is always taken at the entry age y. Changing from NC_t to NC_{t+1} creates a new deficit given by:

$$\Delta UAL(t+1) = \frac{\Delta NC_{t+1}}{_{x+t+1-y}E_y} \ddot{a}_{y,\overline{x+t+1-y}|}.$$

The method therefore leads to an accumulation of deficit amortization. On the other hand, it is easy to show that, when successive deficits are amortized in constant installments over the period remaining until retirement age, the entry age method is identical to the individual level premium method. In practice, the logic of the method leads to amortizing deficits over a shorter period (e.g. 5 years), which therefore leads to a higher level of funding at the beginning of the career.

Finally, it should be noted that the total actuarial liability has the same form as in individual level premium (part of this liability being this time unfunded through the UAL term):

$$AL(t) = \frac{N}{40} 50\% S(t-1)\ddot{a}_{60} \cdot {_{60-x_0-t}E_{x_0+t}} - NC_{t-1}\ddot{a}_{x_0+t,\overline{60-x_0-t}|}.$$

10.9 Pension Gap

When an individual affiliated to a pension scheme changes from one employer to another during his career and is affiliated to a new pension scheme, different issues arise regarding the overlapping of pension plans. Three main concerns need to be addressed:

10.9 Pension Gap

- loss of rights: some pension schemes require a minimum period of affiliation (e.g. 5 years). In the event of early departure before the end of this probationary period, the member loses all his rights; everything happens as if he had never been a member,
- portability: when the affiliated retains his rights, he can in some schemes transfer the reserves accumulated to the new scheme. The reserves thus become "portable",
- the pension gap: whether or not there is portability, a change of employer can result in a loss for the affiliated compared to the situation where he would not have changed employers, even if the two pension schemes are identical.

This phenomenon, called the "pension gap", does not occur in Defined Contribution systems: the accumulated savings simply add up. However, in Defined Benefit schemes, the problem of defining the liabilities arises. How to integrate the benefits of the first plan into the second plan? How to revalue the benefits of the first plan?

Example 10.1 helps to illustrate this problem.

Example 10.1 Let us assume a career from 20 to 60 years old with a change of employer at age 40. It is assumed that both employers have the same pension formula corresponding to a pension of 50% of the final salary for a 40-year career:

$$\text{pension annuity} = \frac{N}{40} 50\% S.$$

We use the unit credit cost method for the funding of the two schemes. Without change of employer, the benefit obtained at retirement age can be symbolized by the following rectangle in the time/salary axes (Fig. 10.3).

What happens if there is a change of employer? The new employer will no longer revalue the parts of the career worked at another employer. So this time we will obtain the schema of Fig. 10.4.

E_1 is the portion funded by the first employer at the time of departure:

$$AL_{20} = \frac{20}{40} 50\% S_{19} \cdot {}_{20}E_{40}$$

to finance an obligation at the age of 60

$$E_1 = \frac{20}{40} 50\% S_{19}.$$

(continued)

Example 10.1 (continued)
E_2 is the portion funded by the second employer giving a commitment

$$E_2 = \frac{20}{40} 50\% S_{39}.$$

E_3 is the pension gap, i.e. the missing portion compared to total obligation

$$E_3 = \frac{40}{40} 50\% S_{39} - E_1 - E_2$$

$$= 50\% S_{39} - \frac{20}{40} 50\% S_{19} - \frac{20}{40} 50\% S_{39}$$

$$= \frac{20}{40} 50\% (S_{39} - S_{19}).$$

Generally speaking, if there is a departure after n years, the pension gap is given by

$$PG = \frac{n}{40} 50\% (S_{39} - S_{n-1}).$$

The pension gap can be measured in relation to the total obligation:

$$PG_n^\% = \frac{PG}{E} = \frac{\frac{n}{40} 50\% (S_{39} - S_{n-1})}{\frac{40}{40} 50\% S_{39}} = \frac{n}{40} \left(1 - \frac{S_{n-1}}{S_{39}}\right).$$

If salaries follow a geometric progression with a common ratio g (i.e. $S_t = S_0(1+g)^t$):

$$PG_n^\% = \frac{n}{40} \left(1 - \frac{S_0(1+g)^{n-1}}{S_0(1+g)^{40-1}}\right) = \frac{n}{40} \left(1 - \frac{1}{(1+g)^{40-n}}\right).$$

Table 10.2 gives, for a salary growth of 5%, the pension gap percentage according to the time of change of employer. The pension gap is small at the beginning of the career since the lost revaluation applies to a few years of service (term in $n/40$). It becomes small again at the end of the career because the difference in salaries becomes small (bell-shaped curve).

10.10 Early Retirement and Actuarial Anticipation

Fig. 10.3 Pension annuity without change of employer

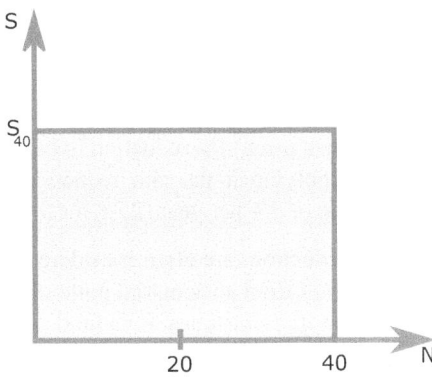

Fig. 10.4 Pension annuity with change of employer (pension gap)

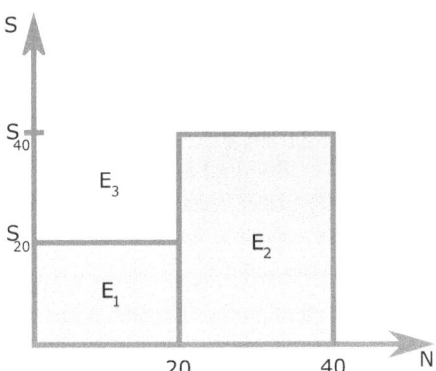

Table 10.2 Pension gap percentage as a function of n, the time of change of employer, as obtained in Example 10.1

n	5	10	20	25	30	37
$PG_n^\%$	10%	19%	31%	32%	29%	13%

10.10 Early Retirement and Actuarial Anticipation

Another form of mobility is to retire early and receive a pension benefit directly.

The impact in financial terms is considerable, since there is both a reduction in the number of contributions and an increase in the benefit period. Statutory pension systems generally organize penalty mechanisms, with reduction coefficients that reduce the pension paid. These reductions, often considered excessive by beneficiaries, are however totally insufficient from an actuarial point of view. This has already been discussed in Sect. 6.2 in a Pay-As-You-Go context. Let us now examine this phenomenon in an individual funded method.

As an example, consider a scheme that grants a retirement pension of 50% of final salary at age 65 for a full career period of 45 years. In the event of early retirement, a linear reduction in benefits is often applied:

- either by simply playing with the number of years:
  - leave at 65: annuity $= \frac{45}{45} 50\% S$,
  - leave at 60: annuity $= \frac{40}{45} 50\% S$,

  which means an 11% reduction in benefits,
- or by applying a flat-rate reduction per year of anticipation. E.g., apply a reduction of 5% per year (25% for 5 years of anticipation).

These reductions are often considered excessive by the beneficiary. However, are they sufficient from an actuarial point of view?

Example 10.2 illustrates, for a funded method, the size of the penalties that would technically have to be applied to be neutral in terms of the cost of the pension plan.

Example 10.2 Assumptions:

- unit pension from age 65,
- funding from age 20 to 65 using the Unit Credit Cost method,
- no mortality until age 65,
- annuity price: certain annuity up to the fixed age of death, which is set at 80,
- discount rate: 4%.

The calculation of the annuity earned at age 60 is then given by

- reserve established at age 60:

$$AL_{60} = \frac{40}{45} \left(\frac{1}{1,04}\right)^5 \ddot{a}_{65},$$

- conversion of this reserve into an annuity at age 60:

$$AL_{60} = k \; \ddot{a}_{60},$$

where k will be the reduction coefficient to be applied.

Equating these two amounts gives the value of k:

$$k = \frac{40}{45} \left(\frac{1}{1,04}\right)^5 \frac{\ddot{a}_{65}}{\ddot{a}_{60}}.$$

Generally speaking, if n years are anticipated, the coefficient becomes

$$k(n) = \frac{45-n}{45} \left(\frac{1}{1,04}\right)^n \frac{\ddot{a}_{65}}{\ddot{a}_{65-n}}.$$

(continued)

10.10 Early Retirement and Actuarial Anticipation

Example 10.2 (continued)
So for a five-year look ahead and a discount rate of 4%, we have

$$k(5) = \frac{40}{45} \left(\frac{1}{1,04}\right)^5 \cdot \frac{\ddot{a}_{65}}{\ddot{a}_{60}}$$

$$= \frac{40}{45} \left(\frac{1}{1,04}\right)^5 \left(\frac{1-v^{15}}{1-v^{20}}\right)$$

$$= 0.5971.$$

i.e. a reduction of 40%, which is far greater than the reductions traditionally encountered.

The size of this penalty is related to the juxtaposition of three reduction effects induced by the coefficient $k(n)$:

- The first term, $\frac{45-n}{45}$, illustrates the fact that some contributions are missing,
- The second term $\left(\frac{1}{1,04}\right)^n$ illustrates that the pension liabilities must be available n years earlier than expected,
- The third term $\frac{\ddot{a}_{65}}{\ddot{a}_{65-n}}$ illustrates that more pension arrears will have to be paid than expected.

Chapter 11
Collective Funding Methods

Abstract This chapter is devoted to the analysis of collective funding methods. After recalling their fundamental characteristics, we describe the leveling methods (based on the same philosophy as the methods presented in Chap. 5 in the context of Pay-As-You-Go) and the pure methods of collective funding.

11.1 General Principles of Collective Funding

As in Chap. 10, we consider a Defined Benefit pension scheme, organized within the second pillar, but this time collectively funded.

There is therefore no longer an individualization of the elements of reserves and contributions but a global vision at the level of the whole group of affiliates. Funding is carried out in a spirit of solidarity within this group, without seeking a balance for each individual.

The main objective of collective funding methods is to provide overall funding for the scheme that respects a certain stability over time. For example, it is desired that the contribution rate, defined as the ratio between the total contributions to be paid each year and the total payroll, should be as constant as possible over time. This consideration leads to the fact that the main collective funding methods used in practice are projected type methods.

Within these collective techniques we can distinguish two major philosophies: leveling methods, and pure collective funding methods.

The leveling methods consist of working in two stages. First the forecast costs of the plan over a defined time horizon are calculated using an individual funding method. In a second stage the overall costs computed this way are leveled. The method is therefore similar to that of the leveling and equalization funds seen in Chap. 5. The underlying element this time is no longer Pay-As-You-Go contributions, but individually funded contributions.

Pure collective funding methods, on the contrary, do not rely on an individual method but directly project the whole group of affiliates. Within this family, the distinction between the major methods is mainly due to the way in which the initial past service at the creation of the plan is funded.

220   11 Collective Funding Methods

Table 11.1 Overview of pure collective funding methods

	Projected
Stable amounts	Aggregate Cost (AGC)
Stable amounts with explicit back service amortization	Attained Age Normal (back services valued in unit credit) (AAN)
	Frozen Initial Liability (back services valued in entry age) (FIL)

As with individual funding, some methods fund these initial past services separately. Table 11.1 summarizes the different situations.

In order to illustrate these various techniques, we will consider the standard pension scheme described in Sect. 10.1, which provides from age 60 a pension corresponding to 50% of the last annual working salary for a full career of 40 years.

11.2 Leveling Techniques

The leveling techniques are based on a two-floor pension scheme funding architecture. Firstly, one applies a given individual funding method with individual accounts. Secondly, one levels the overall costs resulting from floor 1 and finances a leveling fund, also known as a financing fund, or buffer fund.

The calculation procedure consists of three steps:

1. application of the chosen method of individual funding for the first year,
2. projection over a fixed number of years of this individual method using

    - financial assumptions:
        - wage growth,
        - discount rate,
    - demographic assumptions:
        - exits (deaths, resignations, retirements),
        - entries (possible replacements),

3. leveling of these costs.

The following example describes the procedure for our standard pension plan:

$$\text{pension} = \frac{N}{40} 50\% S^{\text{final}} \quad \text{at 60 years old.}$$

Let $\mathcal{J} = \{1, 2, \ldots, K\}$ denote all affiliates present at $t = 0$. We assume that back service (provided before $t = 0$) is not taken into account in the scheme. The unit

11.2 Leveling Techniques

credit cost method is chosen as example of individual underlying method. We denote by x_j the initial age at $t = 0$ of the individual j and $S_j(0)$ his initial salary.

First Step: Calculation of the First Year's Individual Normal Costs

Denoting by $NC_j(0)$ the normal costs for affiliate j at $t = 0$, we obtain in unit credit (cf. Eq. (10.2))

$$NC_j(0) = \frac{1}{40} 50\% S_j(0) \cdot \ddot{a}_{60} \cdot {}_{60-x_j}E_{x_j}.$$

By summing up all the members, we get the total individual funding cost:

$$NC(0) = \sum_{j \in \mathcal{J}} NC_j(0).$$

Second Step: Projection Over T Years (Usually T Between 10 and 20 Years)

The future annual cost is projected for each individual present at $t = 0$. The projection stops when the individual reaches retirement age:

$$NC_j^*(t) = \begin{cases} \left[\frac{1}{40} 50\% S_j^*(t) + \frac{t}{40} 50\% \left(S_j^*(t) - S_j^*(t-1) \right) \right] \\ \qquad \cdot \ddot{a}_{60} \cdot {}_{60-x_j-t}E_{x_j+t} & \text{if } x_j + t < 60 \\ 0 & \text{if } x_j + t \geq 60 \end{cases}$$

(it is assumed here that there are no replacements).

For the projected salary $S_j(t)$, we can for example use a geometric progression:

$$S_j^*(t) = S_j(0) \cdot (1 + g)^t.$$

Third Step: Leveling Out Over the T Years

The variable normal cost vector from the second step is replaced by stable funding as a percentage of total salary. This takes into account the survival probabilities of the affiliates.

For an initial leveling fund of zero ($F(0) = 0$), the leveling rate τ will be the solution to the actuarial equivalence equation

$$\sum_{t=0}^{T-1} \sum_{j \in \mathcal{J}} NC_j^*(t) \cdot \frac{1}{(1+i)^t} \cdot \frac{\ell_{x_j+t}}{\ell_{x_j}} = \sum_{t=0}^{T-1} \sum_{j \in \mathcal{J}} \tau \cdot S_j^*(t) \cdot \frac{1}{(1+i)^t} \cdot \frac{\ell_{x_j+t}}{\ell_{x_j}},$$

where ℓ_x is a given life table. The rate τ depends neither on time (requirement of temporal stability), nor on individuals (collective funding stage).

So the solution is

$$\tau = \frac{\sum_{t=0}^{T-1} \sum_{j \in \mathcal{J}} NC_j^*(t) \cdot \frac{1}{(1+i)^t} \cdot \frac{\ell_{x_j+t}}{\ell_{x_j}}}{\sum_{t=0}^{T-1} \sum_{j \in \mathcal{J}} S_j^*(t) \cdot \frac{1}{(1+i)^t} \cdot \frac{\ell_{x_j+t}}{\ell_{x_j}}}, \quad (11.1)$$

which is sometimes represented as follows:

$$\tau = \frac{PVFNC}{PVFS},$$

where PVFNC is the discounted value of future normal costs, and PVFS is the discounted value of future salaries.

Assuming a non-zero initial fund $F(0) > 0$, it is sufficient to subtract it from the PVFNC:

$$\tau = \frac{PVFNC - F(0)}{PVFS}.$$

Once the contribution rate has been fixed, the total contribution (i.e. the cost to the plan) can be calculated as

$$C(0) = \tau \cdot \sum_{j \in \mathcal{J}} S_j(0).$$

This $C(0)$ amount is compared to the $NC(0)$ amount required to fund individual contracts:

- If $C(0) < NC(0)$, the method cannot be used,
- if $C(0) > NC(0)$, the balance $C(0) - NC(0)$ goes into the leveling fund:

$$F(0^+) = C(0) - NC(0).$$

The following year, before funding, the fund becomes

$$F(1^-) = F(0^+) \cdot (1+i) = (C(0) - NC(0))(1+i)$$

and after funding at $t = 1$

$$F(1^+) = F(1^-) + C(1) - NC(1).$$

11.3 Aggregate Cost

Generally speaking, we have

$$C(t) = \tau \cdot \sum_{j \in \mathcal{J}} S_j(t)$$

$$F(t) = F\left(t - 1^+\right) \cdot (1 + i)$$

$$F\left(t^+\right) = F(t) + C(t) - NC(t).$$

If reality follows the assumptions, the F fund is nil at the end of the period.

The advantages of the method are that it combines the characteristics of individual and collective funding. As in individual funding, there is clear evidence of reserves on an individual by individual basis ("transparency of acquired rights"). As in collective funding, there is a search for collective stability of the scheme funding.

Nevertheless, this method has drawbacks: it requires both double calculation (calculation of individual normal costs and a collective rate), and double management (individual contracts and leveling funds).

11.3 Aggregate Cost

The aggregate cost method is a real collective funding method aimed at obtaining stable funding for a given population as a percentage of the total salary over the lifetime of affiliates.

It has two differences from the leveling method. On the one hand, there are no longer two levels but only one level, the collective level. On the other hand, the projection is done over the lifetime of each affiliate and not over a fixed period for all.

Let us look again at the plan providing a retirement pension equal to

$$\text{pension} = \frac{N}{40} 50\% S^{\text{final}} \quad \text{at 60 years old}$$

and let us denote by $\mathcal{J} = 1, 2, \ldots, K$ the affiliates present at $t = 0$.

For each individual, two projections are made:

- a projection of the benefits:

$$R_j^*(0) = \text{projected benefit of individual } j, \text{ estimated at } t = 0$$

$$= \frac{N_j}{40} 50\% S_j(0) \cdot (1 + g)^{60 - x_j - 1},$$

- a projection of salaries:

$$S_j^*(t) = \text{projected salary at } t \text{ of individual } j$$
$$= S_j(0) \cdot (1+g)^t,$$

Two aggregates are then computed:
- the benefit aggregate

$$PVFB = \sum_{j \in \mathcal{J}} R_j^*(0) \cdot \ddot{a}_{60} \cdot {}_{60-x_j}E_{x_j},$$

- the salary aggregate

$$PVFS = \sum_{j \in \mathcal{J}} \sum_{t=0}^{60-x_j-1} S_j^*(t) \cdot {}_t E_{x_j}.$$

Finally, the contribution rate is calculated using an actuarial equivalence between the benefit aggregate and the contribution aggregate:

$$\sum_{j \in \mathcal{J}} R_j^*(0) \cdot \ddot{a}_{60} \cdot {}_{60-x_j}E_{x_j} = \sum_{j \in \mathcal{J}} \sum_{t=0}^{60-x_j-1} \tau \cdot S_j^*(t) \cdot {}_t E_{x_j}$$

and so

$$\tau = \frac{\sum_{j \in \mathcal{J}} R_j^*(0) \cdot \ddot{a}_{60} \cdot {}_{60-x_j}E_{x_j}}{\sum_{j \in \mathcal{J}} \sum_{t=0}^{60-x_j-1} S_j^*(t) \cdot {}_t E_{x_j}} = \frac{PVFB}{PVFS}.$$

It is interesting to compare this formula for the contribution rate with Formula (11.1) obtained in the leveling method. The denominators both represent a sum of discounted values of future salaries, but the order of the summation signs is reversed. In leveling one works over a fixed period of time, then one observes the individuals present during this period. In aggregate cost, we work with a generation of individuals, then we observe each individual until retirement age.

If the initial $F(0)$ fund is non-zero, we have just to withdraw it from the PVFB:

$$\tau = \frac{PVFB - F(0)}{PVFS}.$$

11.3 Aggregate Cost

Once the rate is calculated, the total contribution of the first year is

$$C(0) = NC(0) = \tau \cdot \sum_{j \in \mathcal{J}} S_j(0).$$

This contribution goes into the collective fund (there are no longer any individual contracts or accounts). This collective fund receives interest and is used to pay benefits directly.

Generally speaking, the collective fund obeys the recurrence relationship

$$F(t) = [F(t-1) + C(t-1)] \cdot (1 + i_t) - B(t),$$

where $C(t-1)$ is the contribution paid to the fund, i_t is the observed rate of return of the fund, and $B(t)$ is the total benefits paid by the fund.

If reality follows the assumptions exactly, the rate τ can remain constant. In practice, there will be gaps and the rate is regularly recalculated by the formula

$$\tau(t) = \frac{PVFB(t) - F(t)}{PVFS(t)},$$

where

- $F(t)$ is the actual asset of the fund at time t,
- $PVFS(t)$ is the discounted value at time t of future salaries (paid after time t), which is given by

$$PVFS(t) = \sum_{j \in \mathcal{J}} \sum_{s=0}^{60-x_j-t-1} S_j^*(s) \cdot {}_s E_{x_j+t},$$

with $S_j^*(s) = S_j(t) \cdot (1+g)^{s-t}$,
- $PVFB(t)$ is the discounted value at time t of future benefits (paid after time t), which is given by

$$PVFB(t) = \sum_{j \in \mathcal{J}} R_j^*(t) \cdot \ddot{a}_{60} \cdot {}_{60-x_j-t} E_{x_j+t},$$

where

$$R_j^*(t) = \frac{N_j}{40} 50\% S_j(t) \cdot (1+g)^{60-x_1-t-1}.$$

As in the case of leveling, the method can sometimes lead to negative values of the fund and must therefore be rejected, as illustrated by Example 11.1.

Example 11.1 Let us consider two affiliates at $t = 0$:

- $j = 1$, whose initial age is $x_1 = 60$ and who entered service at age 25,
- $j = 2$, whose initial age is $x_2 = 40$ and who entered service at age 25.

The scheme provides a lump sum at age 65 equal to $\frac{N}{40} \times 5 S^{\text{final}}$, where N has been counted with back service. We assume a unit salary for all, no inflation or mortality, and zero discounting. The application of the aggregate cost then gives

- the contribution rate $\tau = \frac{2 \cdot 5}{5 + 25} = \frac{1}{3}$,
- the total contribution $C = \frac{2}{3}$.

This contribution remains constant for 5 years. At $t = 5$, the fund has an asset equal to

$$F(5) = 5 \cdot \frac{2}{3} = \frac{10}{3},$$

which is insufficient to pay for the benefit of individual 1, as

$$F\left(5^+\right) = \frac{10}{3} - 5 = -\frac{5}{3}.$$

The method cannot therefore be applied.

Let us note that in terms of final balance, we find at the end of the course the necessary amount

$$F(25) = F\left(5^+\right) + 20 \cdot \frac{1}{3} = -\frac{5}{3} + \frac{20}{3} = \frac{15}{3} = 5,$$

i.e. enough to pay the second capital.

However, the method must be rejected in this case because the fund went through a negative value. This is in fact a simple case of cost inter-generational transfer.

As illustrated in this calculation, the reason for this imbalance is to be found in the funding of the initial back services, recognized by the scheme but not originally provisioned. The aggregate dilutes this funding over the average residual career period of the group and therefore does not induce a sufficient speed of amortization. In order to avoid this type of problem, variants of the aggregate can be developed to deal autonomously with this problem of recognition of initial services.

11.4 Attained Age Normal

In this variant of the aggregate cost, the idea of leveling the projected future costs as well as possible is maintained, while at the same time avoiding problems of insufficient funds.

From this perspective, the funding of initial back services is no longer included in the calculation of the contribution rate but is subject to separated and accelerated amortization. The valuation of these back services is done using the unit credit method and generates additional funding either once or amortized over a fixed number of years.

Example 11.2 illustrates the process.

Example 11.2 Consider the following scheme whose benefit is expressed as an annuity:

$$\text{pension} = \frac{N}{40} 50\% S^{\text{final}} \quad \text{at 60 years old,}$$

where back services are included in N.

Step 1: Initial past service cost. A single global amount is calculated which covers the initial back service for all members:

$$UAL(0) = \sum_{j \in \mathcal{J}} \frac{N_{1j}}{40} 50\% S_j(0) \cdot \ddot{a}_{60} \cdot {}_{60-x_j}E_{x_j}, \tag{11.2}$$

where N_{1j} is the length of service for individual j between entry into service and creation of the scheme. This amount is initially unfunded.

Step 2: Calculation of the contribution rate for future services. An aggregate rate is calculated by ignoring this cost:

$$\tau = \frac{PVFB - UAL(0)}{PVFS}.$$

The normal cost is given as in aggregate:

$$NC(0) = \tau \cdot \sum_{j \in \mathcal{J}} S_j(0).$$

(continued)

Example 11.2 (continued)
Step 3: Additional funding must be provided to amortize the $UAL(0)$ debt. For example, a constant amortization over m years can be used:

$$ADJ(0) = \frac{UAL(0)}{\ddot{a}_{\overline{m}|}}.$$

The total contribution is then

$$C(0) = NC(0) + ADJ(0)$$

$$= \tau \cdot \sum_{j \in \mathcal{J}} S_j(0) + \frac{UAL(0)}{\ddot{a}_{\overline{m}|}}.$$

11.5 Frozen Initial Liability

The method is similar to attained age normal. The initial back service is no longer included in the calculation of the aggregate rate. These back services are valued using the entry age method. The only difference with respect to Attained Age is the way to compute the Unfunded liabilities $UAL(0)$.

In this case and on the basis of the above example, the value of the initial unfunded commitment $UAL(0)$ (cf. Eq. (11.2)) becomes by application of the entry age logic

$$UAL(0) = \sum_{j \in \mathcal{J}} \frac{N_j}{40} 50\% S_j(0) \cdot \ddot{a}_{60} \cdot \frac{_{60-y_j}E_{y_j}}{\ddot{a}_{y_j,\overline{60-y_j}|}} \frac{\ddot{a}_{y_j,\overline{x_j-y_j}|}}{_{x_j-y_j}E_{y_j}},$$

where y_j denotes the age at entry for affiliate j, x_j represents their age at affiliation, N_j represents the total length of service between entry and retirement, and $S_j(0)$ represents the initial salary (at affiliation).

This amount represents the sum of the individual reserves that would have been accumulated in the scheme, if there had been for each participant a leveled contribution from age at entry to membership at $t = 0$ based on known benefits at membership.

Chapter 12
Actuarial Gains and Losses

Abstract This chapter presents various techniques for amortizing actuarial gains and losses arising from differences between the assumptions used and the actual situation. After recalling the main sources of difference and presenting standard amortization techniques, we show how some of the major individual and collective funding methods react.

12.1 Origin of Actuarial Gains and Losses

The individual or collective fully funded methods presented in Chaps. 10 and 11 are based on various assumptions regarding future changes in the environment: salary growth rate, discount rate, mortality trends. Even if the actuary takes great care to use accurate estimators of these quantities, ex post differences will occur between expected and actual parameter values:

- The difference between the discount rate and the real rate of return on assets.
- The difference between the expected rate of salary growth and actual increases.
- The difference between the life table and the actual mortality.

In addition, the actuary may also have to modify certain projection parameters during the life of the scheme in view of the actual evolution of the corresponding indices. Finally, the benefits provided for in the scheme may also be modified at a certain point in time following a reworking of the plan (for example, upward revision of the retirement benefit).

These different elements cause an imbalance between the assets and liabilities of the scheme that needs to be addressed. Two situations can be distinguished in this context depending on whether the cause of the imbalance is the assets or the liabilities of the pension scheme:

- asset gap: imbalance linked to a different evolution of the assets value compared to the assumptions used in the actuarial liabilities. In this case the estimate of the actuarial liabilities is not modified but an unforeseen gap, positive or negative, occurs with respect to the assets. Taking the basic functions of funding, the

$AL(t)$ value remains unchanged, but the F fund value takes on a $F^*(t)$ value different from the expected $F(t)$ value. These differences are due, for example, to a different ex post evolution of the rate of return on assets,
- liability gap: imbalance related to an instantaneous change in the actuarial liability due either to a change in actuarial projection assumptions or a revision of the pension scheme. This time the value of the fund corresponds to its expected value $F(t)$. On the other hand, the actuarial liability changes from an expected value $AL(t)$ to a new value $AL^*(t)$.

Obviously, these two sources of difference can be cumulative. Suppose for example that rates of return differ widely from the initial assumption, and there is therefore an asset gap. At the same time, this significant difference, considered as persistent by the actuary between assumptions and reality, leads to a revision of the discount rate assumption for the future.

This asset-liability gap creates an unfunded accrued liability:

$$UAL(t) = AL(t) - F(t).$$

When this unfunded liability is positive (loss situation), this gap must be amortized using an actuarial method. On the other hand, when the unfunded liability is negative (gain situation), the surplus can either be maintained for prudential reasons, or the future contributions to the scheme can be revised downwards. At the extreme, there may be a temporary "pause of contributions": the surplus is such that the scheme has not to be funded for a certain period by new contributions.

The purpose of this chapter is to show how some of the major funding methods described in Chaps. 10 and 11 respond to these gaps. The following three main methods are discussed here: the unit credit method, the individual level premium method, and the aggregate method.

12.2 Amortization in Unit Credit Cost

We are interested in a standard pension scheme providing a pension equal to half of the last working salary. The assumption is that there is no recognition of initial past service. It is assumed that at time t the actuarial liability and the fund are in equilibrium and given by Relationship (10.1):

$$AL(t) = F(t) = \frac{t}{40} 50\% S(t-1) \cdot \ddot{a}_{60} \cdot {}_{60-x}E_x.$$

The contribution to be paid at time t in the absence of actuarial gains or losses is given by Relationship (10.2):

$$NC_t = \left[\frac{1}{40} 50\% S(t) + \frac{t}{40} 50\% (S(t) - S(t-1)) \right] \ddot{a}_{60} \cdot {}_{60-x}E_x.$$

12.2 Amortization in Unit Credit Cost

When the assumptions follow reality, this amount effectively allows us to go from $AL(t)$ to $AL(t+1)$ and thus represents in this assumption the contribution of the year.

Indeed we can write the following recurrence relation:

$$AL(t+1) = \frac{AL(t) + NC_{t+1}}{{}_1E_x} = \frac{F(t) + NC_{t+1}}{{}_1E_x} = F(t+1), \qquad (12.1)$$

expressing that the movement of liabilities from t to $t+1$ is funded from the asset side by income and from the liability side by the payment of normal cost.

12.2.1 Asset Gap

Let us first consider the case of an asset gap. In this case, the actuarial liability formulas are not changed and the contribution payable must still allow for a shift from the $AL(t)$ level to the $AL(t+1)$ level.

In reality, the pure endowment evolution must be calculated using real elements (effective rate of return between t and $t+1$ and real survival rate). Formula (12.1) therefore implicitly assumes that these bases are identical to those used in the actuarial liability:

$$ {}_1E_x = \frac{1}{1+i} \, {}_1p_x. \qquad (12.2)$$

In the event of an asset difference, this pure endowment differs from that expressed in technical bases:

$$ {}_1E_x^* = \frac{1}{1+j} \, {}_1p_x^* \qquad (12.3)$$

because the effective rate of return j differs from the actuarial rate i and/or because actual survival differs from the table of mortality. In this case, Relationship (12.1) is no longer verified.

In terms of assets, the payment of the normal contribution NC_t as well as the actual evolution of interest and mortality between times t and $t+1$ allows the fund to go from the amount $F(t)$ to an amount denoted $F^*(t+1)$ given by

$$\begin{aligned} F^*(t+1) &= \frac{AL(t) + NC_t}{{}_1E_x^*} \\ &= \frac{AL(t) + NC_t}{{}_1E_x} - (AL(t) + NC_t)\left(\frac{1}{{}_1E_x} - \frac{1}{{}_1E_x^*}\right) \\ &= AL(t+1) - UAL(t+1). \end{aligned}$$

The amount $UAL(t+1)$ represents an asset shortfall in relation to the actuarial liability. The logic of the unit credit method, which aims to have a reserve corresponding to past service at all times, encourages the funding of this amount in lump sum. In this case, the contribution to be made at time t becomes

$$C(t+1) = NC_{t+1} + UAL(t+1) = NC_{t+1} + (AL(t) + NC_t)\left(\frac{1}{_1E_x} - \frac{1}{_1E_x^*}\right). \quad (12.4)$$

Sometimes, rather than being funded immediately, this deficit can also be amortized over a fixed number of years.

It should be remembered that, in Formula (12.4), the UAL amount may, depending on the case, be positive (actuarial loss—additional funding to be made) or negative (actuarial gain—reduction in funding). In particular, in the case of a large gain (UAL greater than NC), the scheme is in a "pause of contributions" situation.

12.2.2 Liability Gap

Let us now consider the situation of a liability difference resulting, for example, from a change in some of the actuarial assumptions used to determine the actuarial liability. Assuming that there is no asset difference, Relationship (12.1) (and therefore the payment of the normal cost) leads to an actuarial liability corresponding to the former technical bases:

$$AL(t+1) = \frac{AL(t) + NC_t}{_1E_x} = \frac{F(t) + NC_t}{_1E_x} = F(t+1)$$

$$= \frac{t+1}{40} 50\% S(t) \cdot \ddot{a}_{60} \cdot _{60-x-1}E_{x+1}.$$

Now the change in actuarial assumptions at time $t+1$ leads to a new actuarial liability target denoted $AL^*(t+1)$ and given by

$$AL^*(t+1) = \frac{t+1}{40} 50\% S(t) \cdot \ddot{a}_{60}^* \cdot _{60-x-1}E_{x+1}^*,$$

the actuarial elements $\ddot{a}_{60}^*$ and $_{60-x-1}E_{x+1}^*$ being calculated using the new assumptions (discount rate and/or life table).

In this case there is therefore a deficit given by

$$UAL(t+1) = AL^*(t+1) - AL(t+1)$$

$$= AL(t+1)\left(\frac{\ddot{a}_{60}^* \cdot _{60-x-1}E_{x+1}^*}{\ddot{a}_{60} \cdot _{60-x-1}E_{x+1}} - 1\right).$$

12.2.3 Scheme Modification

Finally, a change in the definition of the guarantees offered to members can be seen as a source of gap.

As an example, suppose that at time $t+1$ the plan moves from a target of 50% of the last salary to a target of 60% of the last salary with full catch-up of the years already worked (the new formula applies retroactively from enrollment). Then, there is also a liability gap, with the actuarial liability moving from the level

$$AL(t+1) = \frac{t+1}{40} 50\% S(t) \cdot \ddot{a}_{60} \cdot {}_{60-x-1}E_{x+1}$$

to the new level

$$AL^*(t+1) = \frac{t+1}{40} 60\% S(t) \cdot \ddot{a}_{60} \cdot {}_{60-x-1}E_{x+1}.$$

So the deficit in this case is simply

$$UAL(t+1) = \frac{t+1}{40} 10\% S(t) \cdot \ddot{a}_{60} \cdot {}_{60-x-1}E_{x+1}.$$

This deficit can again be funded either immediately or spread over a fixed number of years.

Note finally that in case of cumulation of different sources of deviation it is enough to add the various corrections seen above.

12.3 Amortization in Individual Level Premium

Let us now examine how the individual level premium method reacts to actuarial gains and losses.

It is assumed that up to time t there has been no source of gains or losses. Then between times t and $t+1$ a difference is introduced either in assets or liabilities. At time t, the normal cost of the method can be written (cf. (10.4))

$$NC_t = NC_{t-1} + \Delta NC_t, \tag{12.5}$$

$$\Delta NC_t = \left[\frac{N}{40} 50\% \Delta S(t) \ddot{a}_{60}\right] \cdot \frac{{}_{60-x_0-t}E_{x_0+t}}{\ddot{a}_{x_0+t,\overline{60-x_0-t|}}}.$$

Actuarial liabilities can be calculated using a retrospective formula expressing the reserve as the accumulated sum of past payments prior to time t. In the absence of an actuarial difference, this amount is also equal to the fund:

$$AL(t) = F(t) = \sum_{j=0}^{t-1} \frac{NC_j}{{}_{t-j}E_{x_0+j}}.$$

If there is no difference between the times t and $t+1$, the normal cost at time $t+1$ can be expressed as a function of this actuarial liability in an alternative way to Formula (12.5). The normal cost at time $t+1$ is the amount that makes it possible to fund the balance of the unfunded liabilities in level premiums.

$$NC_{t+1} = \left[\frac{N}{40} 50\% S(t+1) \ddot{a}_{60} \cdot {}_{60-x_0-t-1}E_{x_0+t+1} \right.$$

$$\left. - \frac{AL(t) + NC_t}{{}_1E_{x_0+t}} \right] \frac{1}{\ddot{a}_{x_0+t+1,\overline{60-x_0-t-1|}}}. \quad (12.6)$$

Actuarial liabilities follow the same recurrence formula (12.1) as in unit credit.

12.3.1 Asset Gap

Let us now consider the case of an asset gap between times t and $t+1$. We can then define, as in unit credit, a deficit of assets compared to the actuarial liabilities:

$$UAL(t+1) = (AL(t) + NC_t) \left(\frac{1}{{}_1E_{x_0+t}} - \frac{1}{{}_1E^*_{x_0+t}} \right),$$

where pure endowments are given by (12.2) and (12.3).

If the basic philosophy of the individual level premium method of leveling charges by constant payments is to be maintained, this deficit should be amortized over the rest of the career. In this case, the amortization of the implicit deficit will be included in the normal cost. The normal cost then becomes

$$NC^*_{t+1} = \left[\frac{N}{40} 50\% S(t+1) a_{60} \cdot {}_{60-x_0-t-1}E_{x_0+t+1} \right.$$

$$\left. - \frac{AL(t) + NC_t}{{}_1E_{x_0+t}} + UAL(t+1) \right] \frac{1}{\ddot{a}_{x_0+t+1,\overline{60-x_0-t-1|}}}$$

12.3 Amortization in Individual Level Premium

or

$$NC^*_{t+1} = \left[\frac{N}{40} 50\% S(t+1) a_{60} \cdot {}_{60-x_0-t-1}E_{x_0+t+1}\right.$$
$$\left. - \frac{AL(t) + NC_t}{{}_1 E^*_{x_0+t}}\right] \frac{1}{\ddot{a}_{x_0+t+1, \overline{60-x_0-t-1|}}},$$

which is similar to the formula without deviation (12.6), except that the pure endowment between the instants t and $t+1$ is no longer calculated in the technical bases but using the actual parameters.

Another approach is to use an explicit amortization and to fund this deficit separately, either immediately or over a fixed period of time.

In the case of a single amortization, the total contribution to be paid at time $t+1$ is given as in unit credit by

$$C(t+1) = NC_{t+1} + UAL(t+1) = NC_{t+1} + (AL(t) + NC_t)\left(\frac{1}{{}_1 E_{x_0+t}} - \frac{1}{{}_1 E^*_{x_0+t}}\right).$$

12.3.2 Liability Gap

Let us now consider the case of a liability difference resulting from a change in actuarial assumptions at time $t+1$. In the standard case where one wishes to define a leveled normal cost in the new technical bases, it is sufficient to adapt, in Formula (12.6), the actuarial elements which are subject to revision:

$$NC^*_{t+1} = \left[\frac{N}{40} 50\% S(t+1) \ddot{a}^*_{60} \cdot {}_{60-x_0-t-1}E^*_{x_0+t+1}\right.$$
$$\left. - \frac{AL(t) + NC_t}{{}_1 E_{x_0+t}}\right] \frac{1}{\ddot{a}^*_{x_0+t+1, \overline{60-x_0-t-1|}}}.$$

12.3.3 Scheme Modification

The same logic can be used when the liability difference is caused not by a change in assumptions but by a change in the pension scheme. For instance, in the hypothesis already encountered where the benefit increases from 50 to 60% with retroactive effect, the new normal cost is given by

$$NC^*_{t+1} = \left[\frac{N}{40} 60\% S(t+1) \ddot{a}_{60} \cdot {}_{60-x_0-t-1}E_{x_0+t+1} - \frac{AL(t) + NC_t}{{}_1 E_{x_0+j}} \right] \ddot{a}_{\overline{x_0+t+1,60-x_0-t-1|}} \cdot \frac{1}{}.$$

12.4 Amortization in Aggregate Cost

The study of actuarial gains and losses in the aggregate cost method takes a particular form. Due to its structure, this method does not explicitly highlight a difference between actuarial liabilities and the value of assets. Amortization is therefore by definition always implicit.

Indeed, Formula (11.3) for calculating the scheme's contribution rate already mixes liabilities (the discounted value of future benefits) and the value of the fund:

$$\tau(t) = \frac{PVFB(t) - F(t)}{PVFS(t)}.$$

Whether asset or liability gaps are involved, the application of aggregate cost will not lead to the calculation of a deficit to be amortized as in the two other methods seen above. It will simply lead to the calculation of a new contribution rate that takes into account these changes in the environment. It could therefore be said that the method by definition automatically amortizes any difference observed during the life of the plan over the residual funding period. This is obviously an advantage of the method. But it is also a danger since it never really measures past errors, except through large changes in contribution levels. This is one of the reasons why this method is sometimes rejected in practice.

More specifically, if there is a difference in assets between times t and $t+1$, the fund moves from an expected level $F(t+1)$ to a new level $F^*(t+1)$ (e.g. due to a different return than expected).

In this case, the contribution rate initially given by

$$\tau(t+1) = \frac{PVFB(t+1) - F(t+1)}{PVFS(t+1)}$$

becomes

$$\tau^*(t+1) = \frac{PVFB(t+1) - F^*(t+1)}{PVFS(t+1)}.$$

12.5 Stochastic Amortization Model

Similarly, in the event of a difference in liabilities, whether caused by a change in actuarial assumptions or by a change in scheme, the contribution rate becomes

$$\tau^*(t+1) = \frac{PVFB^*(t+1) - F(t+1)}{PVFS^*(t+1)},$$

where $PVFB^*(t+1)$ and $PVFS^*(t+1)$ are calculated using the new assumptions.

12.5 Stochastic Amortization Model

12.5.1 Introduction

This section is based on the model developed by Dufresne [35]. We consider a Defined Benefit and individual fully funded pension scheme. The normal cost NC and the actuarial liability AL are therefore calculated for each member of the pension scheme. Then we sum up the total for the entire group. So we can write for the actual total contribution

$$C(t) = \sum_{x=\text{ind}} NC(x, t) + ADJ(t).$$

To determine the value of the ADJ adjustment, the spread method will be used here, which consists of calculating the total unfunded liability (UAL) each year and amortizing it over a given period, denoted in the following by M (M being an integer greater than or equal to 1). One of the objectives of this section is to see if there are better choices to be made for this parameter M. In other words, is it interesting to amortize the deficits over a longer or shorter period of time?

According to the spread method, we have

$$ADJ(t) = \frac{1}{\ddot{a}_{\overline{M}|}} \cdot UAL(t) = \frac{1}{\ddot{a}_{\overline{M}|}} \cdot (AL(t) - F(t)),$$

where

$$F(t) = \text{value of the fund (assets)},$$

$$AL(t) = \sum_{x=\text{ind}} AL(t, x).$$

One can then write a relationship linking the fund to the contributions, which constitutes a negative "feedback" relationship:

$$\begin{aligned} C(t) &= \sum_{x=\text{ind}} NC(x,t) + \frac{1}{\ddot{a}_{\overline{M}|}}(AL(t) - F(t)) \\ &= \sum_{x=\text{ind}} NC(x,t) + k \cdot (AL(t) - F(t)), \end{aligned} \quad (12.7)$$

with

$$k = \frac{1}{\ddot{a}_{\overline{M}|}} = \text{amortization factor.}$$

In the following we will model the quantities F and C by stochastic processes whose source of randomness lies only in the rates of return on the assets.

12.5.2 Stochastic Model for the Fund Value

We use a stochastic investment model and look at its impact on the evolution of the fund and contributions, linked by Eq. (12.7). On the other hand, the stationarity of all other phenomena (demography, inflation, etc.) will be assumed. The model developed here is based on the following eight assumptions:

(A1) All ex ante actuarial assumptions are verified ex post except for the interest rate.
(A2) The population affiliated to the scheme is stationary.
(A3) We work with deflated values. We therefore ignore inflation and the interest rates used are in fact real interest rates. Pensions are assumed to be indexed like wages.
(A4) For the ex ante actuarial valuation, a fixed interest rate (discount rate) is used, denoted i below.
(A5) The real rate of return for the period $(t, t+1)$ is a random variable denoted $i(t+1)$. Returns are therefore a series of random variables.
(A6) All these rates have the same average: $E(i(t)) = i^*$. When $i = i^*$, the actuarial valuation is said to be correct on average.
(A7) Contributions and benefits are assumed to be paid annually in advance. Valuations are performed annually (discrete time model).
(A8) The initial available value of the fund is known and is given by the positive value $F(0) = F_0$.

12.5 Stochastic Amortization Model

The consequence of these assumptions is that the following quantities are constant over time:

$$NC = \sum_x NC(x, t) = \text{total normal cost},$$

$$AL = \sum_x AL(t, x) = \text{total actuarial liability},$$

$$B = \sum_x B(x, t) = \text{total paid benefits},$$

$$S = \sum_x S(x, t) = \text{total salaries}.$$

On the other hand, an identity relationship can be obtained between these quantities. In general, we have the following recurrence relationship on actuarial liabilities (fundamental recurrence relation of the prospective reserve):

$$AL(t+1) = (1+i) \cdot (AL(t) + NC(t) - B(t)).$$

These elements remaining constant in this model, we get

$$AL = (1+i) \cdot (AL + NC - B).$$

In particular, the following relationship linking benefits to contributions and interest cost on liabilities can be derived:

$$B = NC + \frac{i}{(1+i)} AL. \qquad (12.8)$$

Similar relationships can be obtained for the asset side:

$$F(t+1) = (1 + i(t+1)) \cdot (F(t) + C(t) - B). \qquad (12.9)$$

The contributions being defined by the feedback relationship resulting from (12.7) are equal to

$$C(t) = NC + k \cdot (AL - F(t)).$$

In order to study the behavior of the F fund and C contribution processes (including their expectations and dispersions), a model for the $i(t)$ rate-of-return process should be chosen. Here we will add the following two assumptions about this process:

(A9) Random variables $i(t)$ are assumed to be independent and identically distributed.

(A10) The first two moments of these rates exist and are given by

$$E(i(t)) = i^* = i,$$
$$\text{Var}(i(t)) = \sigma^2.$$

12.5.3 Asymptotic Behavior of the Fund Value and the Contributions

Our objective in this section is, given assumptions (A1) to (A10), to get, if they exist, the asymptotic values of the expectations and variances of the fund and contributions:

$$EF(\infty) = \lim_{t \to \infty} E(F(t)), \qquad EC(\infty) = \lim_{t \to \infty} E(C(t)),$$
$$\text{Var} F(\infty) = \lim_{t \to \infty} \text{Var}(F(t)), \qquad \text{Var} C(\infty) = \lim_{t \to \infty} \text{Var}(C(t)).$$

This will allow us to see if the pension fund is converging when the scheme reaches maturity.

12.5.3.1 First Order Moments

Let us first calculate the fund's expectation for a finite time. Given (12.9), we have

$$F(t+1) = (1 + i(t+1)) \cdot (F(t) + C(t) - B)$$
$$= (1 + i(t+1)) \cdot ((1-k)F(t) + NC - B + k \cdot AL).$$

Which can recursively be written as

$$F(t+1) = w(t+1)[q \cdot F(t) + u],$$

where

$$w(t+1) = \frac{1 + i(t+1)}{1 + i}$$
$$q = (1+i) \cdot (1-k)$$
$$u = (1+i) \cdot (NC - B + k \cdot AL).$$

12.5 Stochastic Amortization Model

Taking into account Assumption (A9) of independence of the rates of return and Assumption (A10), the expectation is then the solution of the recursive scheme

$$EF(t+1) = q \cdot EF(t) + u, \tag{12.10}$$

whose solution is given by

$$EF(t) = q^t F_0 + u \frac{1-q^t}{1-q}. \tag{12.11}$$

As far as the contributions are concerned, we know from (12.7) that

$$C(t) = NC + k \cdot (AL - F(t))$$

and therefore

$$EC(t) = NC + k \cdot (AL - EF(t))$$

or explicitly

$$EC(t) = NC + k \cdot \left(AL - \left(F(0) \cdot q^t + u \frac{1-q^t}{1-q} \right) \right). \tag{12.12}$$

We then obtain the following asymptotic property which tells us that whatever their initial value, the fund and contribution processes converge towards their actuarial basis.

Proposition 12.1 *If $0 < q < 1$, then*

$$\lim_{t \to \infty} EF(t) = AL,$$

$$\lim_{t \to \infty} EC(t) = NC.$$

Proof If $0 < q < 1$, Relationship (12.11) allows us to take the limit:

$$\lim_{t \to \infty} EF(t) = \frac{u}{1-q}.$$

Let us calculate this limit by taking into account Relation (12.8), which can be written

$$NC - B = -\frac{i}{1+i} \cdot AL$$

and then we get

$$\frac{u}{1-q} = \frac{(1+i)(NC - B + k \cdot AL)}{1 - (1+i)(1-k)}$$

$$= \frac{(1+i)}{1 - (1+i)(1-k)} \cdot \left(AL \frac{k(1+i) - i}{(1+i)}\right)$$

$$= AL.$$

The contribution relationship then follows directly from (12.12). □

Remark 12.1 What does the $0 < q < 1$ condition mean in practice? This number q is explicitly given by

$$q = (1+i)(1-k) = (1+i)\left(1 - \frac{1}{\ddot{a}_{\overline{M}|}}\right)$$

$$= \frac{\left(\ddot{a}_{\overline{M}|} - 1\right)(1+i)}{\ddot{a}_{\overline{M}|}}$$

$$= \frac{\ddot{a}_{\overline{M-1}|}}{\ddot{a}_{\overline{M}|}} \quad \text{(for } M > 1\text{)}.$$

For a valuation rate $i > 0$, this ratio is such that

$$0 < \frac{\ddot{a}_{\overline{M-1}|}}{\ddot{a}_{\overline{M}|}} < 1.$$

We thus obtain the following corollary, which is a sufficient condition for asymptotic convergence.

Corollary 12.1 *If the amortization period is strictly greater than 1 (i.e. if $M > 1$) and if the valuation rate (which is also the average return on assets) is strictly positive ($i > 0$), the mathematical expectation of the fund tends asymptotically towards the actuarial liability.*

Remark 12.2 Let us look in detail at the case where $M = 1$ (i.e. immediate amortization of any deficit or surplus identified). We then have $k = 1$ and $q = 0$. Relationship (12.11) then becomes

$$EF(t) = u$$

$$= (1+i)(NC - B + AL)$$

$$= (1+i)\left(NC - \left(NC + \frac{i}{1+i}AL\right) + AL\right)$$

$$= AL.$$

12.5 Stochastic Amortization Model

In this case, the fund's expectation already reaches its asymptotic level AL at any finite time $t > 0$.

Remark 12.3 The assumption of equality between the actuarial valuation rate i and the average rate of return on assets i^* (see assumption (A10)) obviously plays a central role in these different results. Let us examine what happens when these two rates differ. The relation (12.10) then becomes

$$EF(t+1) = \frac{1+i^*}{1+i} \cdot (q \cdot EF(t) + u)$$

$$= q^* \cdot EF(t) + u^*,$$

where

$$q^* = q \frac{1+i^*}{1+i} \quad \text{and} \quad u^* = u \cdot \frac{1+i^*}{1+i}.$$

As a result, if $q* < 1$, we have an asymptotic convergence to

$$EF(\infty) = \frac{u^*}{1-q^*} = \frac{u}{1-(1+i^*)(1-k)} \cdot \frac{1+i^*}{1+i}$$

$$= AL \cdot \frac{1+i^*}{1+i} \cdot \frac{k(1+i)-i}{k(1+i^*)-i^*}$$

$$= AL \cdot f(i, i^*).$$

The equilibrium level AL is therefore perturbed by a function of i and i^* that satisfies the following properties:

1. $f(i, i) = 1$,
2. $f(i, i^*) > 1$ when $i^* > i$ and $f(i, i^*) < 1$ when $i^* < i$.

This is a corollary of the growth of the function of a real variable:

$$h(x) = \frac{1+x}{(k(1+x)-x)},$$

whose first derivative is given by

$$h'(x) = \frac{1}{(k(1+x)-x)^2} > 0.$$

Under the convergence condition ($q* < 1$), we can logically deduce that when the actuarial rate is lower than the rate of return on assets (i.e. when the performance of the assets is better than expected), the fund's expectation converges towards a higher level than that of the actuarial liabilities (and vice versa).

12.5.3.2 Second Order Moments

Let us now calculate the dispersions of the fund and of the contributions around their average value calculated above. It is assumed throughout this subsection that $i = i^*$.

Proposition 12.2 *The variance of the fund satisfies the recursive relation*

$$\mathrm{Var}\, F(t+1) = a\,\mathrm{Var}\, F(t) + b(\mathrm{E} F(t+1))^2, \tag{12.13}$$

where

$$a = (1-k)^2\left((1+i)^2 + \sigma^2\right) = (1-k)^2 y^2,$$
$$b = \frac{\sigma^2}{(1+i)^2}.$$

In particular, we explicitly obtain the variance of the fund and contributions:

$$\mathrm{Var}\, F(t) = b \cdot \sum_{j=1}^{t} a^{t-j} (\mathrm{E} F(j))^2, \tag{12.14}$$

$$\mathrm{Var}\, C(t) = k^2 \cdot \mathrm{Var}\, F(t).$$

Proof We start from the recurrence relationship of the fund:

$$F(t+1) = w(t+1)[q F(t) + r].$$

Given the assumption of independence of annual financial returns, we obtain the moment of order 2 (independence between the variables $w(t+1)$ and $F(t)$):

$$\mathrm{E}\left(F^2(t+1)\right) = \mathrm{E}\left(w^2(t+1)\right)\mathrm{E}\left((q F(t) + r)^2\right).$$

On the other hand

$$\mathrm{E}\left(w^2(t+1)\right) = \frac{1}{(1+i)^2}\mathrm{E}\left((1+i(t+1))^2\right) = \frac{1}{(1+i)^2}\left((1+i)^2 + \sigma^2\right)$$

$$= 1 + \frac{\sigma^2}{(1+i)^2}.$$

And by the relationship of (12.10)

$$\mathrm{E}\left((q F(t) + r)^2\right) = \mathrm{E}\left(\left((q(F(t) - \mathrm{E} F(t)) + (q\mathrm{E} F(t) + r)\right)^2\right)$$
$$= q^2 \mathrm{Var}\, F(t) + (\mathrm{E} F(t+1))^2.$$

12.5 Stochastic Amortization Model

So we have

$$E\left(F^2(t+1)\right) = \left(1 + \frac{\sigma^2}{(1+i)^2}\right) \cdot \left(q^2 \cdot \operatorname{Var} F(t) + (EF(t+1))^2\right).$$

Finally the desired variance becomes

$$\operatorname{Var} F(t+1) = E\left(F^2(t+1)\right) - (EF(t+1))^2$$

$$= \left(1 + \frac{\sigma^2}{(1+i)^2}\right) \cdot q^2 \cdot \operatorname{Var} F(t) + \frac{\sigma^2}{(1+i)^2}(EF(t+1))^2$$

$$= a \cdot \operatorname{Var} F(t) + b \cdot (EF(t+1))^2,$$

where

$$b = \frac{\sigma^2}{(1+i)^2}$$

$$a = q^2 \cdot \left(1 + \frac{\sigma^2}{(1+i)^2}\right)$$

$$= (1+i)^2 (1-k)^2 \left(1 + \frac{\sigma^2}{(1+i)^2}\right)$$

$$= (1-k)^2 \left((1+i)^2 + \sigma^2\right).$$

This proves Relation (12.13), of which Relation (12.14) is an immediate corollary. □

These results again allow us to obtain asymptotic values under certain conditions on the parameters.

Proposition 12.3 *If $a < 1$, then asymptotic fund and contribution variances exist and are given by*

$$\lim_{t \to \infty} \operatorname{Var} F(t) = b \cdot \frac{(AL)^2}{1-a}$$
$$\lim_{t \to \infty} \operatorname{Var} C(t) = b \cdot k^2 \cdot \frac{(AL)^2}{1-a}. \qquad (12.15)$$

Proof This results directly from Relation (12.14) and from the fact that $EF(t)$ converges to AL. □

Remark 12.4 What does the $a < 1$ condition mean in practice? For this, let us calculate this number a and transform the condition on a into a condition on k and therefore on M (i.e. the amortization period):

$$a = (1-k)^2 \left((1+i)^2 + \sigma^2\right).$$

The condition $a < 1$ is therefore equivalent to

$$k > 1 - \frac{1}{\left((1+i)^2 + \sigma^2\right)^{1/2}}.$$

By replacing k according to rate i, we obtain

$$\frac{i}{\left(1 - v^M\right)(1+i)} > 1 - \frac{1}{\left((1+i)^2 + \sigma^2\right)^{1/2}}.$$

Isolating the value of M we get

$$M < M^* = \frac{1}{\ln(1+i)} \ln\left(\frac{(1+i)(1+b)^{0.5} - 1}{(1+b)^{0.5} - 1}\right), \tag{12.16}$$

where

$$b = \frac{\sigma^2}{(1+i)^2}.$$

Contrary to the case of the expectation, where the convergence condition on q was trivially verified (cf. Remark 12.1), we now have a real condition on the parameters. We thus obtain the following corollary, which tells us that one should not amortize over too long a period (thus not reject the charges of the past in a too distant future by choosing an overly long amortization period) if one does not want an explosive dispersion of the processes.

Corollary 12.2 *When the valuation rate is strictly positive and when the amortization period M is less than the M^* limit value defined in Eq. (12.16), then the fund and the contributions have finite asymptotic variances.*

Table 12.1 gives the limit value M^* for various combinations of the average rate and its volatility.

Table 12.1 Limit value M^*

	$i = 1\%$	$i = 5\%$
$\sigma = 5\%$	223	78
$\sigma = 15\%$	66	37
$\sigma = 20\%$	42	28

12.5.4 Optimal Amortization Period

We have just obtained an upper limit value for the amortization period M if an explosion of the dispersion is to be avoided. However, below this limit, are there better choices than others? Can an optimal amortization period be established according to some criterion to be defined?

One criterion that comes to mind spontaneously would be to jointly minimize the asymptotic variability of both the fund (interest of affiliates) and the contributions (interest of the sponsor).

Let us study the limit values of the variance of the fund and the contributions as a function of the amortization parameter k (and thus indirectly as a function of the amortization period M). Taking into account (12.15), we get for the variance of the fund

$$\operatorname{Var} F(\infty) = b \cdot \frac{AL^2}{(1-a)}$$

$$= \frac{\sigma^2}{(1+i)^2}(AL)^2 f(k),$$

where

$$f(k) = \frac{1}{\left(1 - (1-k)^2 \cdot \left((1+i)^2 + \sigma^2\right)\right)} = \frac{1}{\left(1 - (1-k)^2 y^2\right)}$$

and

$$y^2 = (1+i)^2 + \sigma^2.$$

Note that we find the boundary condition seen in Remark 12.1. Indeed, for this variance to be finite and positive, $f(k)$ must be positive, i.e.

$$k > 1 - \frac{1}{y},$$

which is the limit condition found on k.

Similarly for contributions, we have

$$\operatorname{Var} C(\infty) = b \cdot k^2 \frac{AL^2}{(1-a)}$$

$$= \frac{\sigma^2}{(1+i)^2}(AL)^2 g(k),$$

where

$$g(k) = \frac{k^2}{\left(1 - (1-k)^2 \cdot y^2\right)}.$$

Therefore we have to study the behavior of functions f and g as a function of k (thus M). First, for the variance of the fund, we have the following property.

Proposition 12.4 *The asymptotic variance of the fund is an increasing function of the amortization period.*

Proof Since the function k is a decreasing function of M, it is sufficient to show that the function $f(k)$ is a decreasing function of k. Let us calculate its first derivative, which is negative (k being smaller than 1):

$$f'(k) = \frac{-2(1-k)y^2}{\left(1-(1-k)^2 y^2\right)^2} < 0.$$

□

The situation is somewhat different for contributions because, unlike the $f(k)$ function, the function $g(k)$ is not monotonous. Indeed we have the following property.

Proposition 12.5 *The asymptotic variance of contributions is first a decreasing function of M for $M < M^{**}$ and then becomes an increasing function of M for $M > M^{**}$. The value of M^{**} is given by*

$$M^{**} = -\frac{1}{\ln(1+i)} \ln\left(\frac{vy^2 - 1}{y^2 - 1}\right),$$

with $v = 1/(1+i)$.

Proof Let us now analyze the behavior of the function $g(k)$. Taking the derivative, we get

$$g'(k) = \frac{\left(1-(1-k)^2 y^2\right) \cdot 2k - k^2 \cdot 2(1-k)y^2}{\left(1-(1-k)^2 y^2\right)^2}$$

$$= \frac{2ky^2 \left(k - (1 - 1/y^2)\right)}{\left(1-(1-k)^2 y^2\right)^2} = \frac{2ky^2}{\left(1-(1-k)^2 y^2\right)^2} h(k).$$

The linear function $h(k)$ thus changes sign for the value k^* given by

$$k^* = 1 - 1/y^2.$$

12.5 Stochastic Amortization Model

The derivative is positive for $k > k^*$ (increasing function) and negative for $k < k^*$ (decreasing function). For $k = k^*$, we have a minimum value of the variance. The conclusion is therefore the same when moving to the variable M. The limit value M^{**} corresponding to the value k^* can then be determined easily by solving the equation

$$\frac{1}{\ddot{a}_{\overline{M^{**}|}}} = 1 - 1/y^2.$$

This equation in M can be written as

$$\frac{1}{\ddot{a}_{\overline{M|}}} = \left((1+i)\frac{1-v^M}{i}\right)^{-1} = \beta,$$

whose solution is

$$M = \frac{1}{\ln v} \ln\left(1 - \frac{i}{\beta(1+i)}\right).$$

By substituting β for its value, we finally obtain

$$M^{**} = -\frac{1}{\ln(1+i)} \ln\left(1 - \frac{iv}{(1-1/y^2)}\right)$$

$$= -\frac{1}{\ln(1+i)} \ln\left(\frac{vy-1}{y-1}\right).$$

□

Propositions 12.4 and 12.5 together allow us to define a zone of optimality for the parameter M.

Corollary 12.3 *The optimal amortization period lies in the interval* $1 \leq M \leq M^{**}$.

Proof In this interval, we are in a region where if we increase M we simultaneously have an increase in the variance of the fund (disadvantage!) but a decrease in the variance of the contributions (advantage). On the other hand, to the right of M^{**}, both variances become increasing and there is no longer any interest in increasing the amortization period. □

Table 12.2 gives the values of M^{**} for various combinations of the average rate and its volatility. In practice many pension funds use an amortization period between 10 and 15 years, which seems to be consistent with these values.

Table 12.2 Limit value M^{**}

	$i = 1\%$	$i = 5\%$
$\sigma = 5\%$	60	14
$\sigma = 15\%$	28	11
$\sigma = 20\%$	19	10

Remark 12.5 Let us check that the optimal bound M^{**} is lower than the maximum M^* defined previously. To do this, simply check that the values of k corresponding to these two values of M are in reverse order, i.e. $k^* > k$. Indeed we have

$$k^* = 1 - 1/y^2 > k = 1 - 1/y,$$

as

$$y^2 = (1+i)^2 + \sigma^2 > 1.$$

Appendix A
Technical Appendix

Abstract In this appendix, we introduce the basic actuarial notation and concepts, which are used throughout the book.

A.1 The Two Basic Tools of Actuarial Science

Consider the following problem as a first example: a 20-year-old individual wishes to receive an annual income of €10,000 at the end of each year starting from retirement at age 65 and continuing for as long as he/she lives. What is the annual premium to be paid during the working years? We will attempt to introduce the necessary tools one by one to perform this calculation.

In a very general sense, actuarial computations are concerned with future financial amounts that need to be evaluated today. They are based on two fundamental tools, the two actuarial "atoms":

- *Discounting*: €1 in 30 years is not worth €1 today, considering the interest that can be earned on this €1 if it is invested for 30 years. Financial discounting is therefore based on an interest rate that is supposed to reflect the average future returns on an investment.
- *Life tables*: In life insurance or pension theory, future payments are generally contingent on a random event related to human life. For example, a pension capital is paid at retirement only if the beneficiary is still alive at that time. An annuity is payable as long as its beneficiary is alive. The actuary incorporates these survival probabilities through a tool called a "life table".

We detail below the basic principles of these two fundamental components of actuarial calculations. They are the two building blocks, the salt and pepper, of both life insurance and pension theory.

A.2 Compound Interest and Capitalization

An initial sum of money is assumed to be invested in a bank. We are interested in what this amount will become in the future, considering the interest that will be earned from the bank. We work with compound interest: the interest earned progressively is left in the account and in turn earns interest.

A.2.1 Capitalization over One Year

Problem A capital of €1000 is invested at $t = 0$ at an annual interest rate of 8%; what is the amount obtained after one year?

Solution We compute, denoting by $C(t)$ the value of the capital at time t,

$$C(1) = \text{initial capital} + \text{interest}$$
$$= C(0) + C(0) \cdot 0.08$$
$$= C(0) \cdot (1 + 0.08)$$
$$= €1000 \cdot 1.08$$
$$= €1080.$$

Generalization If an initial capital $C(0)$ is invested at an interest rate i, the amount obtained after one year is

$$C(1) = C(0) \cdot (1 + i).$$

A.2.2 Capitalization over Several Years

Problem A capital of €1000 is invested at $t = 0$ at an annual interest rate of 8%; what is the amount obtained after 5 years?

Solution If we are at $t = 2$, the capital can be obtained from the capital of the previous year:

$$C(2) = C(1) \cdot 1.08$$

and we can then use the decomposition of $C(1) = C(0) \cdot 1.08$ obtained in previous problem:

$$C(2) = C(1) \cdot 1.08 = (C(0) \cdot 1.08) \cdot 1.08 = C(0) \cdot 1.08^2.$$

A Technical Appendix

We then iterate this process to obtain $C(5)$:

$$\begin{aligned} C(5) &= C(4) \cdot 1.08 = (C(3) \cdot 1.08) \cdot 1.08 \\ &= C(3) \cdot 1.08^2 = (C(2) \cdot 1.08) \cdot 1.08^2 \\ &= C(2) \cdot 1.08^3 = (C(1) \cdot 1.08) \cdot 1.08^3 \\ &= C(1) \cdot 1.08^4 = (C(0) \cdot 1.08) \cdot 1.08^4 \\ &= C(0) \cdot 1.08^5 = {€}1000 \cdot 1.469 = {€}1469. \end{aligned}$$

Generalization If an initial capital $C(0)$ is invested at an interest rate i for n years, the amount obtained is

$$C(n) = C(0) \cdot (1+i)^n.$$

Figure A.1 shows the evolution of the capital discussed in the previous problem.

This method of capitalization is called *compound interest*. It is important to note that we take the nth power of $(1+i)$, as opposed to the *simple interest* method, where the rate is multiplied by n:

$$\text{compound interest:} \quad (1+i)^n \neq 1 + i \cdot n \quad \text{: simple interest.}$$

Thus, in the previous problem,

$$1.08^5 = 1.469 \neq 1.4 = 1 + 0.08 \cdot 5.$$

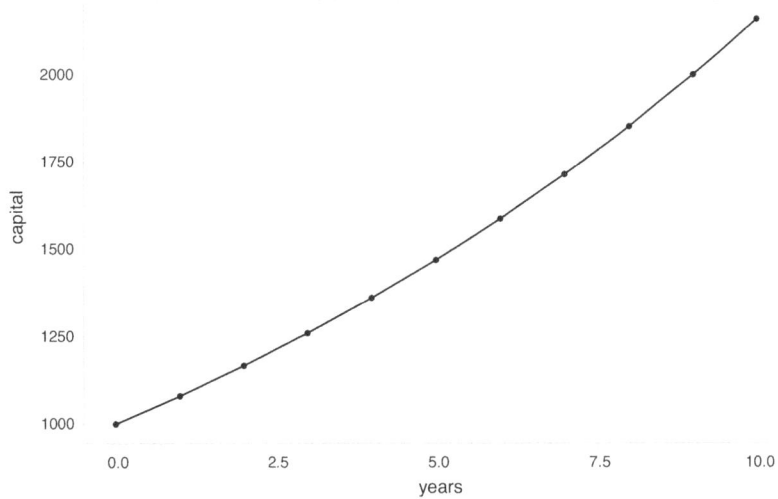

Fig. A.1 Evolution of the capital when capitalized over several years

The simple interest method is used in finance in particular situations not considered here.

A.2.3 Required Capitalization Time

Problem An amount of €1000 is invested at an annual rate of 5%; after how long will the capital be worth €2000?

Solution We use the formula given in the generalization of the previous problem, but this time the unknown is the duration n:

$$C(n) = C(0) \cdot (1+i)^n \quad \text{gives here} \quad €2000 = €1000 \cdot 1.05^n,$$

whose solution is given by

$$n = \log_{1.05}\left(\frac{€2000}{€1000}\right) = \log_{1.05}(2) = 14.2067.$$

Generalization The time required for a capital C invested at a rate i to reach the value D is equal to

$$n = \log_{(1+i)} \frac{D}{C}.$$

A.2.4 Capitalization of an Annuity

Problem At the beginning of each year for 5 years, a capital of €1000 is invested at an annual interest rate of 8%; what is obtained after 5 years?

Solution The future value (FV) is equal to the sum of 5 terms:

$$FV(5) = €1000 \cdot 1.08^5 + €1000 \cdot 1.08^4 + €1000 \cdot 1.08^3 + €1000 \cdot 1.08^2$$
$$+ €1000 \cdot 1.08$$
$$= €1000 \cdot \left(1.08^5 + 1.08^4 + 1.08^3 + 1.08^2 + 1.08\right)$$
$$= €1000 \cdot 6.33593$$
$$= €6335.93.$$

A Technical Appendix

There is a simpler formula to perform this calculation:

$$FV(5) = €1000 \cdot \frac{1.08^5 - 1}{0.08} \cdot 1.08 = €1000 \cdot 6.33593.$$

Generalization If a capital C is invested at the beginning of each year for n years at an interest rate i, the amount obtained at the end is given by

$$FV(n) = C \cdot (1+i)^n + C \cdot (1+i)^{n-1} + \cdots + C \cdot (1+i)^2 + C \cdot (1+i)$$

$$= C \cdot \left((1+i)^n + (1+i)^{n-1} + \cdots + (1+i)^2 + (1+i)\right)$$

$$= C \cdot \sum_{j=1}^{n} (1+i)^j$$

$$= C \cdot \frac{(1+i)^n - 1}{i} \cdot (1+i)$$

$$= C \cdot \ddot{s}_{\overline{n}|},$$

where we have used the actuarial notation $\ddot{s}_{\overline{n}|}$, in which the umlaut indicates that the payments are made at the beginning of the year.

If the payments are not made at the beginning of the year (annuity due) but at the end of the year (ordinary annuity), the future value is then

$$FV(n) = C \cdot (1+i)^{n-1} + C \cdot (1+i)^{n-2} + \cdots + C \cdot (1+i) + C$$

$$= C \cdot \left((1+i)^{n-1} + (1+i)^{n-2} + \cdots + (1+i) + 1\right)$$

$$= C \cdot \sum_{j=0}^{n-1} (1+i)^j$$

$$= C \cdot \frac{(1+i)^n - 1}{i}$$

$$= C \cdot s_{\overline{n}|},$$

where we have used the actuarial notation $s_{\overline{n}|}$, in which the absence of the umlaut (compared to the previous notation $\ddot{s}_{\overline{n}|}$) indicates that the payments are made at the end of the year.

A.3 Discounting

A.3.1 Discounting over One Year

Problem What amount needs to be paid today (at $t = 0$) to obtain a sum of €1000 in one year if the bank grants us an interest rate of 8%?

Solution Given what we have seen in the previous section, it is necessary to pay a capital C which is the solution to the equation

$$C \cdot 1.08 = €1000,$$

that is

$$C = \frac{€1000}{1.08} = 925.93.$$

This amount, when invested at an interest rate of 8%, indeed produces a sum of €1000:

$$C \cdot 1.08 = \frac{€1000}{1.08} \cdot 1.08 = €1000.$$

Generalization The present value of an amount C to be paid in one year at an interest rate i is given by

$$PV = \frac{C}{1+i} = C \cdot v,$$

where we use the standard notation $v = \frac{1}{(1+i)}$.

A.3.2 Discounting over Several Years

Problem What amount needs to be paid today (at $t = 0$) to obtain a sum of €1000 in 5 years if the bank grants us an interest rate of 8%?

Solution In the same way as for the previous problem, we solve

$$C \cdot 1.08^5 = €1000 \quad \Rightarrow \quad C = \frac{€1000}{1.08^5} = €680.58.$$

A Technical Appendix

Generalization The present value of an amount C to be paid in n years at an interest rate i is given by

$$PV = \frac{C}{(1+i)^n} = C \cdot v^n.$$

A.3.3 Annuity in Present Value

Problem What is the present value at $t = 0$ of a series of investments made at the beginning of each year for 5 years of an amount of €1000 at an annual interest rate of 8%?

Solution As with capitalized annuities, we simply sum the present values:

$$PV = €1000 \cdot 1.08^{-4} + €1000 \cdot 1.08^{-3} + €1000 \cdot 1.08^{-2}$$
$$+ €1000 \cdot 1.08^{-1} + €1000$$
$$= €1000 \cdot \left(1.08^{-4} + 1.08^{-3} + 1.08^{-2} + 1.08^{-1} + 1\right)$$
$$= €1000 \cdot 4.31213$$
$$= €4312.13.$$

Similarly, a more direct formula exists for this calculation:

$$PV = €1000 \cdot \frac{1 - 1.08^{-5}}{0.08} \cdot 1.08 = €1000 \cdot 4.31213.$$

Generalization If an amount C is invested at the beginning of each year for n years at an interest rate i, the present value at the origin is given by:

$$PV = C \cdot \frac{1 - (1+i)^{-n}}{i} \cdot (1+i) = C \cdot \ddot{a}_{\overline{n}|},$$

where we have used the actuarial notation $\ddot{a}_{\overline{n}|}$, in which the umlaut indicates that the payments are made at the beginning of the year.

In the case of payments made at the end of the year (ordinary annuity), the present value is equal to

$$PV = C \cdot \frac{1 - (1+i)^{-n}}{i} = C \cdot a_{\overline{n}|},$$

where the actuarial notation $a_{\overline{n}|}$ is used.

A.4 Life Tables

All the quantities calculated so far are financial values that do not take into account the chances of survival. In insurance and pensions, many amounts are conditional on whether or not the insured lives. For this reason, we need to introduce survival probabilities in addition to the financial functions described above. In actuarial science, the tool used to calculate these survival probabilities is the life table.

A life table starts with a cohort (a group of individuals born in the same year) of 1,000,000 people at birth; each year, the number of survivors from this initial group is tracked. We then define

$$\ell_x = \text{number of survivors at age } x \text{ among } 1{,}000{,}000 \text{ people}.$$

Table A.1 is an excerpt from one of the Belgian life tables (MR table). By convention, the last age in the table (the age at which all individuals die) is denoted ω. In the case of the above table, $\omega = 111$. Figure A.2 shows the evolution of the cohort in the MR table.

Using such a table, we can then calculate various probabilities of survival and death.

Table A.1 Excerpt of the Belgian MR life table

Age	ℓ_x	Survivors
0	ℓ_0	1,000,000
1	ℓ_1	999,415
2	ℓ_2	998,827
...	...	...
20	ℓ_{20}	987,349
21	ℓ_{21}	986,616
...	...	...
64	ℓ_{64}	850,437
65	ℓ_{65}	839,161
66	ℓ_{66}	826,964
67	ℓ_{67}	813,786
...	...	...
85	ℓ_{85}	366,772
...	...	...
100	ℓ_{100}	16,452
...	...	...
110	ℓ_{110}	23
111	ℓ_{111}	0

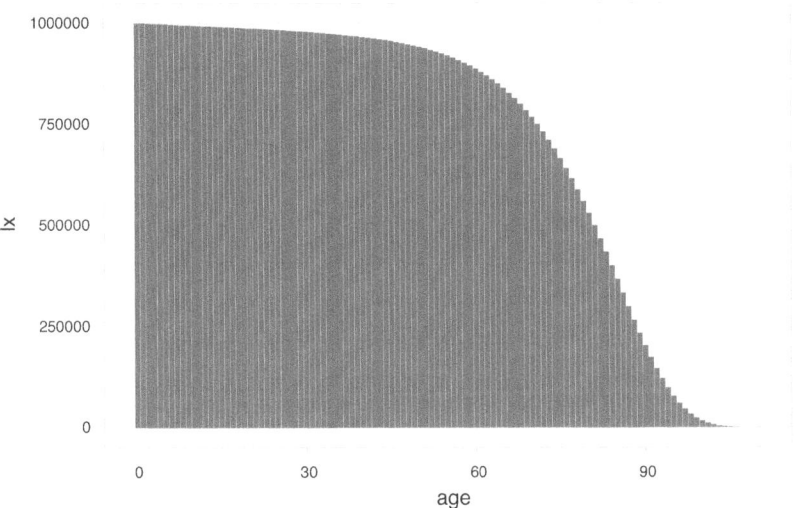

Fig. A.2 Evolution of the MR life table cohort

A.4.1 Probability of Surviving One Year

Problem What is the probability that someone aged 20 today will survive a year from now?

Solution The life table is used to estimate the number of cases favorable to the "survival" event and the number of total cases:

$$p_{20} = \mathbb{P}(\text{survival from age 20 to age 21}) = \frac{\ell_{21}}{\ell_{20}} = \frac{986,616}{987,349} = 0.9993 = 99.93\%.$$

Generalization The probability of survival from age x to age $x+1$ is equal to

$$p_x = \frac{\ell_{x+1}}{\ell_x}.$$

A.4.2 Probability of Death over One Year

Problem What is the probability that someone aged 65 today will die within the year?

Solution The "death" event is the complementary of the "survival" event, and the probability is therefore

$$q_{65} = \mathbb{P}(\text{death at age 65})$$
$$= 1 - \mathbb{P}(\text{survival from age 65 to age 66})$$
$$= 1 - p_{65}$$
$$= 1 - \frac{\ell_{66}}{\ell_{65}}$$
$$= 1 - \frac{826,964}{839,161}$$
$$= 0.0145 = 1.45\%.$$

Generalization The probability of death at age x (before age $x+1$) is

$$q_x = 1 - p_x = 1 - \frac{\ell_{x+1}}{\ell_x} = \frac{\ell_x - \ell_{x+1}}{\ell_x}.$$

Note that $\ell_x - \ell_{x+1}$ (often denoted d_x) is the number of deaths in one year among the population of age x, and that the last expression therefore coincides with the definition of the probability of death as the quotient of the number of favorable cases by the number of total cases.

Figure A.3 shows the q_x obtained from the MR table.

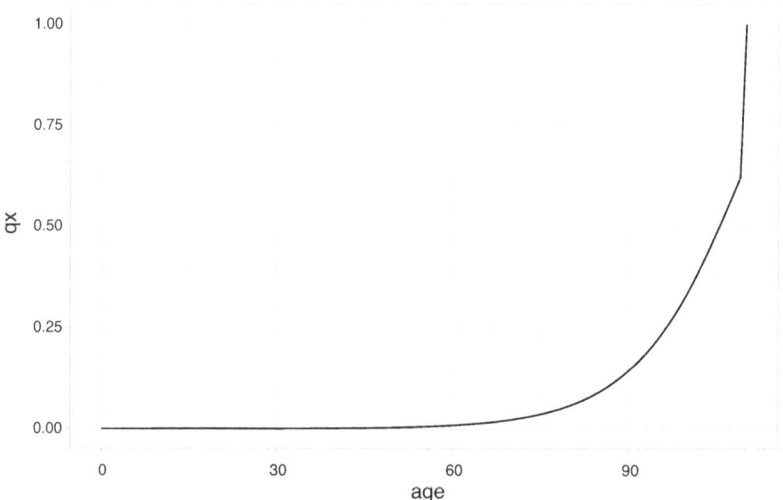

Fig. A.3 Probabilities of death q_x from the MR table

A Technical Appendix

A.4.3 Probability of Surviving Several Years

Problem What is the probability of a 20-year-old surviving to age 65?

Solution The calculation is carried out in the same way as for survival over one year:

$$_{45}p_{20} = \mathbb{P}(\text{survival from age 20 to age 65}) = \frac{\ell_{65}}{\ell_{20}} = \frac{839,161}{987,349} = 0.8499 = 85\%.$$

Generalization The probability of survival for n years of an individual of age x is equal to

$$_np_x = \frac{\ell_{x+n}}{\ell_x}.$$

For $n = 1$, the previous notation is often used: $_1p_x = p_x$.

Note that the probability of survival over n years can be expressed as the product of n probabilities of survival over one year:

$$\prod_{k=0}^{n-1} p_{x+k} = p_x \, p_{x+1} \, p_{x+2} \cdots p_{x+n-1}$$

$$= \frac{\ell_{x+1}}{\ell_x} \frac{\ell_{x+2}}{\ell_{x+1}} \frac{\ell_{x+3}}{\ell_{x+2}} \cdots \frac{\ell_{x+n}}{\ell_{x+n-1}}$$

$$= \frac{\ell_{x+n}}{\ell_x}$$

$$= {_np_x}.$$

A.5 Life Contracts

In the category of "life contracts", we put actuarial instruments that offer payments if the insured is alive.

A.5.1 Pure Endowment

Pure endowment is the basic life operation; it combines the two tools seen above: compound interest and probability of survival. It is the basic atom of life insurance.

The *pure premium* of an insurance contract is the mathematical expectation of what the contract costs the insurer, in present value terms (or, equivalently, what the contract pays the insured). In this way, if the insurer has a large portfolio, the total premiums collected are on average equivalent to what he has to pay out. The commercial premium, which is what the customer actually pays, is the pure premium plus various loadings.

Problem How much would an individual aged 20 have to pay if they wanted to obtain €100,000 at age 65 if they lived to that age, using an interest rate of 3%?

Solution This quantity can be considered as the pure premium of an insurance contract. We thus calculate the expectation of the random variable "cost to the insurer" in present value terms:

$$\mathbb{E}\left(\frac{1}{1.03^{45}} \cdot \text{cost}\right) = \frac{1}{1.03^{45}} \cdot \text{cost(survival)} \cdot \mathbb{P}(\text{survival})$$

$$+ \frac{1}{1.03^{45}} \cdot \underbrace{\text{cost(death)}}_{=0 \text{ (life contract)}} \cdot \mathbb{P}(\text{death})$$

$$= \frac{1}{1.03^{45}} \cdot \text{cost(survival)} \cdot \mathbb{P}(\text{survival})$$

$$= \frac{1}{1.03^{45}} \cdot \text{€}100,000 \cdot \frac{839,161}{987,349}$$

$$= \text{€}22,475.$$

Generalization The price of pure endowment C between ages x and $x+n$ at a rate i is given by

$$C \cdot {}_nE_x = C \cdot \frac{1}{(1+i)^n} \cdot \frac{\ell_{x+n}}{\ell_x} = C \cdot v^n \cdot {}_np_x,$$

where we have used the actuarial notation ${}_nE_x$, which specifies the starting age x and duration of the contract n.

A.5.2 Temporary Life Annuity

A temporary life annuity consists in paying a fixed amount each year (or each month, ...) for a given period, as long as the insured survives. Payment stops automatically at a certain age (for example, during working life until retirement). It can therefore be interpreted as a sum of pure endowments.

Problem An individual aged 20 wishes to pay an annual amount of €1000 at the beginning of each year for as long as he/she lives, until retirement age of 65. What

A Technical Appendix

does he/she have to pay in a single lump sum at age 20 to replace all these payments at an interest rate of 3%?

Solution Since a life annuity corresponds to a sum of pure endowment, its price is equal to

$$€1000 \cdot \ddot{a}_{20,\overline{45|}} = €1000 + \frac{€1000}{1.03} \cdot p_{20} + \frac{€1000}{1.03^2} \cdot {}_2p_{20}$$

$$+ \cdots + \frac{€1000}{1.03^{44}} \cdot {}_{44}p_{20}$$

$$= €1000 \cdot \left(1 + \frac{1}{1.03} \cdot \frac{986,616}{987,349} + \frac{1}{1.03^2} \cdot \frac{985,864}{987,349}\right.$$

$$\left. + \cdots + \frac{1}{1.03^{44}} \cdot \frac{850,437}{987,349}\right)$$

$$= €24,574.36.$$

Generalization The price of an anticipated temporary life annuity at age x, at a rate i, for an amount C and duration n is given by

$$C \cdot \ddot{a}_{x,\overline{n|}} = C + \frac{C}{1+i} \cdot p_x + \frac{C}{(1+i)^2} \cdot {}_2p_x + \cdots + \frac{C}{(1+i)^{n-1}} \cdot {}_{n-1}p_x$$

$$= C \cdot \sum_{k=0}^{n-1} \frac{1}{(1+i)^k} \cdot {}_kp_x = C \cdot \sum_{k=0}^{n-1} v^k \cdot {}_kp_x,$$

where we have used the actuarial notation $\ddot{a}_{x,\overline{n|}}$, which can be seen as an equivalent of $\ddot{a}_{\overline{n|}}$ in which the probability of survival has been taken into account (the reason for specifying the age x).

In the case of an ordinary annuity, the notation $a_{x,\overline{n|}}$ is used (that is, the umlaut is simply dropped).

A.5.3 Lifetime Annuity

A lifetime annuity (also known simply as a life annuity) is an annuity that only stops when the insured dies.

Problem A retiree aged 65 wants to receive a pension of €10,000 at the end of each year for as long as he/she lives. What does he/she have to pay today, given an interest rate of 3%?

Solution The computation is similar to that of the price of a temporary life annuity, except that it is the age ω that corresponds to the end of the payments:

$$€10,000 \cdot a_{65} = \frac{€10,000}{1.03} \cdot p_{65} + \frac{€10,000}{1.03^2} \cdot {}_2p_{65} + \cdots + \frac{€10,000}{1.03^{\omega-66}} \cdot {}_{\omega-66}p_{65}$$

$$= €10,000 \left(\frac{1}{1.03} \cdot \frac{826,964}{839,161} + \frac{1}{1.03^2} \cdot \frac{813,786}{839,161} \right.$$

$$\left. + \cdots + \frac{1}{1.03^{\omega-66}} \cdot \frac{23}{839,161} \right)$$

$$= €128,523.08.$$

Generalization The price of an ordinary lifetime annuity at age x, for an amount C and at a rate i is

$$C \cdot a_x = \frac{C}{1+i} \cdot p_x + \frac{C}{(1+i)^2} \cdot {}_2p_x + \cdots + \frac{C}{(1+i)^{\omega-x-1}} \cdot {}_{\omega-x-1}p_x$$

$$= C \cdot \sum_{k=1}^{\omega-x-1} \frac{1}{(1+i)^k} \cdot {}_kp_x.$$

In pension theory, the quantity a_x is called the "annuity price".

Whenever the annuity is due (i.e. when the payments are to be made at the beginning of the year), the notation for the annuity price is $\ddot{a}_x$.

A.5.4 Application: Level Premium

Problem A 20-year-old individual wishes to obtain an annual income of €10,000 at the end of each year, starting from his retirement at age 65 and for as long as he lives. What is the premium to be paid each year during his working life at an interest rate of 3%?

Solution The computation is divided into four steps:

1. Computation of the required provision at age 65:

$$C = €10,000 \cdot a_{65} = €10,000 \cdot 12.852308 = €128,523.08.$$

2. Present value of benefits at 20 years:

$$V_1 = C \cdot {}_{45}E_{20} = €128,523.08 \cdot 0.22475 = €28,885.57.$$

A   Technical Appendix

3. Present value of premiums at 20 years:

$$V_2 = p \cdot \ddot{a}_{20,\overline{45|}} = p \cdot €24.57436.$$

4. Actuarial equivalence:

$$V_1 = V_2 \quad \Rightarrow \quad €28,885.57 = p \cdot 24.57436$$

$$\Rightarrow \quad p = \frac{€28,885.57}{24.57436} = €1,175.44.$$

Generalization The annual premium p to be paid between age x and age $x + n$ to obtain a life annuity of amount R from age $x + n$ is given by:

$$p = \frac{R \cdot {}_nE_x \cdot a_{x+n}}{\ddot{a}_{x,\overline{n|}}}.$$

Appendix B
Further Reading

1. Aaron, H.: The social insurance paradox. Can. J. Econ. Pol. Sci. **23**, 371–374 (1966)
2. Aaron, H.: Social security reconsidered. National Tax J. **64**(2), 385–414 (2011)
3. Alonso-Garcia, J.: Notional (non-financial) defined contributions: essays on sustainability and pension adequacy. Ph.D. Thesis, Université Catholique de Louvain, 2015
4. Alonso-Garcia, J., Boado-Penas, M.C., Devolder, P.: Adequacy, fairness and sustainability of pay as you go system: defined benefit versus defined contribution. Eur. J. Financ. **24**(13), 1–27 (2017)
5. Alonso-Garcia, J., Boado-Penas, M.C., Devolder, P.: Automatic balancing mechanisms for notional defined contribution accounts in the presence of uncertainty. Scand. Actuar. J. **2018**(2), 85–108 (2018)
6. Anderson, A.W.: Pension Mathematics for Actuaries. Actex Publications, Connecticut (1992)
7. Arnold, S., Boado-Penas, M.C., Godinez-Olivares, H.: Longevity risk in notional defined contribution pension schemes: a solution. Geneva Papers Risk Insurance-Issues Practice Longevity Risks **41**, 24–52 (2015)
8. Auerbach, A.J., Lee, R.: Notional defined contribution pension systems in a stochastic context: design and stability. iN: Brown, J., Liebman, J., Wise, D. (eds.) Social Security Policy in a Changing Environment. University of Chicago Press, Chicago (2006)
9. Barr, N., Diamond, P.A.: Improving Sweden's automatic pension adjustment mechanism. Center Retirement Res. Briefs **11** (2011)
10. Belloni, M., Maccheroni, C.: Actuarial fairness when longevity increases: an evaluation of the Italian Pension system. Geneva Papers Risk Insurance-Issues Practice **38**, 638–674 (2013)
11. Berin, B.N.: The Fundamentals of Pension Mathematics. Society of Actuaries, New York (1986)
12. Bertocchi, M., Schwartz, S.L., Ziemba, W.T.: Optimizing the Aging, Retirement and Pensions Dilemma. Wiley, Hoboken (2009)
13. Billig, A., Ménard, J.: Actuarial balance sheets as a tool to assess the sustainability of social security pension systems. Int. Soc. Secur. Rev. **66**(2), 31–52 (2013)
14. Blake, D.: Pension Economics. Wiley, Hoboken (2006)
15. Blake, D.: Pension Finance. Wiley, Hoboken (2006)
16. Blake, D., Cairns, A., Dowd, K.: Pensionmetrics: Stochastic pension plan design and value at risk during the accumulation phase. Insur. Math. Econ. **29** (2001)
17. Boado-Penas, M.C., Dominguez-Fabian, I., Vidal-Meliá, C.: Notional defined contribution accounts (NDCs): solvency and risk, application to the case of Spain. Int. Soc. Secur. Rev. **60**(4), 105–127 (2007)

18. Boado-Penas, M.C., Vidal-Meliá, C.: Non-financial defined contribution pension schemes: is a survivor dividend necessary to make the system balanced? Appl. Econ. Lett. **21**(4), 242–247 (2014)
19. Börsch-Supan, A.: What are NDC Systems? What do they bring to reform strategies? In: Pension Reform: Issues and Prospects for Non-Financial Defined Contribution (NDC) Schemes. The World Bank, Washington D.C. (2006)
20. Börsch-Supan, A., Reilheld, A., Wilke, C.B.: How to make a Defined Benefit System sustainable. MEA Discussion Papers **37**, 1–36 (2003)
21. Boulier, J.F., Dupré, D.: Gestion financière des fonds de retraite. Economica, Paris (2002)
22. Bowers, N.L., Hickman, J.C., Gerber, H.U., Nesbitt, C.J., Jones, D.A.: Actuarial Mathematics. Society of Actuaries, Illinois (1986)
23. Bozio, A., Piketty, T.: Pour un nouveau système de retraite: des comptes individuels de cotisations financés par répartition. Editions Rue d'Ulm, Paris (2008)
24. Buchanan, J.: Social insurance in a growing economy: a proposal for radical reform. National Tax J. **21**(4), 386–389 (1968)
25. Campbell, J., Viceira, L.: Strategic asset allocation–portfolio choice for long term investors. Oxford University Press, Oxford (2002)
26. Chlon-Dominczak, A., Franco, D., Palmer, E.: The first wave of NDC Reforms: the experiences of Italy, Latvia, Poland and Sweden. Nonfinancial Defined Contribution Pension Schemes in a Changing Pension World, vol. 1. The World Bank, Washington D.C. (2012)
27. Collinson, D.: Actuarial methods and assumptions used in the valuation of retirement benefits in the EU and other European countries. European Actuarial Consultative Group, Oxford (2001)
28. Commission de Réforme des Pensions 2020–2040.: Un contrat social performant et fiable. Technical Report, SPF Sécurité Sociale (2014). https://www.conseilacademiquepensions.be/docs/fr/rapport-062014-fr.pdf
29. Devolder, P., de Valeriola, S.: Between DB and DC: optimal hybrid PAYG pension schemes. Eur. Actuar. J. **9**, 463–482 (2019)
30. Devolder, P., Melis, R.: Optimal mix between pay as you go and funding for pension liabilities in a stochastic framework. ASTIN Bulletin **45**, 551–575 (2015)
31. Devolder, P., Janssen, J., Manca, R.: Stochastic Methods Pension Funds. Wiley, Hoboken (2012)
32. Devolder, P., Hindriks, J., Schokkaert E., Vandenbroucke, F.: Réforme des pensions légales: le système de pension à points. Regards Econ. **130**, 1–7 (2017)
33. Dickson, D., Hardy, M.R., Waters, H.R.: Actuarial Mathematics for Life Contingent Risks. Cambridge University Press, Cambridge (2020)
34. Disney, R.: Notional Accounts as a Pension Reform Strategy: An Evaluation. The World Bank, Washington, D.C. (1999)
35. Dufresne, D.: Moments of pension fund contributions and fund levels when rates of return are random. J. Inst. Actuar. **115**, 535–544 (1988)
36. Dufresne, D.: Mathématiques des Caisses de Retraite. Editions Supremum, Montréal (1994)
37. Dupuis, J.M., El Moudden, C.: Economie des Retraites. Economica, Paris (2002)
38. Estrada, M.A.R., Koutrona, E.: The Basic Manual of Social Security. Theory and Evaluation: An Introduction to Social Security. Lambert Academic Publishers, London (2020)
39. Feraud, L.: Mathématiques et théories actuarielles à l'usage des assurances sur la vie, caisses de retraite et régimes par répartition. Dunod, Paris (1971)
40. Fujisawa, Y., Siu-Hang Li, J.: The impact of the automatic balancing mechanism for the public pension in Japan on the extreme elderly. North Am. Actuar. J. **16**(2), 207–239 (2012)
41. Gannon, F., Legros, F., Touzé, V.: Automatic adjustment mechanisms and budget balancing of pension schemes. Working Papers of the Observatoire Français des Conjonctures Économiques (2013)
42. Gannon, F., Hamayon, S., Legros, F., Touzé, V.: Sustainability of the French first pillar pension scheme (CNAV): assessing automatic balance mechanisms. Aust. J. Actuar. Practice **2**, 33–45 (2014)

B Further Reading

43. Gibrais, V., Adam, A.C.: Le Calcul des Engagements de Retraite Supplémentaire. Economica, Paris (2004)
44. Gollier, J.J.: L'avenir des Retraites. L'Argus, Paris (1987)
45. Gronchi, S., Nisticò, S.: Implementing the NDC theoretical model: a comparison of Italy and Sweden. In: Holzmann, R., Palmer, E. (eds.) Pension Reform Issues and Prospects for Non-Financial Defined Contribution (NDC) Schemes, pp. 493–515. The World Bank, Washington D.C. (2005)
46. Gronchi, S., Nistico, S.: Theoretical foundations of pay as you go defined contribution pension schemes. Metroeconomica **59**(2), 131–159 (2008)
47. Guérin, J.L., Legros, F.: Neutralité actuarielle: un concept élégant mais délicat à mettre en œuvre. Revue d'économie Financiére **68**, 79–90 (2002)
48. Gurtovaya, V., Nistico, S.: The notional defined contribution pension scheme and the German point system: a comparison. German Econ. Rev. **19**(4), 1–18 (2017)
49. Haberman, S.: Autoregressive rates of return and the variability of pension fund contributions and fund levels for a DB pension scheme. Insur. Math. Econ. **14**, 219–240 (1994)
50. Hindriks, J.: Quel Avenir pour nos pensions ? Les grands défis de la réforme des pensions. De Boeck, Brussels (2015)
51. Holzman, R.: The ABCs of Nonfinancial Defined Contribution (NDC) Schemes. Int. Soc. Secur. Rev. **70**(3), 1–23 (2017)
52. Holzman, R., Palmer, E.E.: Pension Reform: Issues and Prospects for non-financial defined contribution (NDC) schemes. The World Bank, Washington D.C. (2006)
53. Holzman, R., Palmer, E.E.: Nonfinancial Defined Contribution Pension Schemes in a Changing Pension World. The World Bank, Washington D.C. (2012)
54. Jagob, J., Sesselmeier, W.: Is there an optimal way out of the pension crisis? An investigation of different approaches. In: Pieters, D. (ed.) Confidence and Changes: Managing Social Protection in the New Millennium, pp. 45–78. Kluwer Law International, Alphen aan den Rijn (2001)
55. Kalfon, P., Peubez, G.: L'actuariat des Engagements Sociaux. Economica, Paris (2004)
56. Kessler, D., Strauss-Kahn, D.: L'épargne et la Retraite. Economica, Paris (1982)
57. Khorasanee, M.Z.: Survey of actuarial practice in the funding of UK defined benefit pension schemes. Actuarial research papers **54**, 1–20 (1994)
58. Knell, M.: How automatic adjustment factors affect the internal rate of return of PAYG pension systems. J. Pension Econ. Financ. **9**, 1–23 (2010)
59. Legros, F.: Notional defined contribution: a comparison of the French and German point system. Economics Papers from the Centre d'études Prospectives et d'informations Internationales **14**, 1–39 (2003)
60. Lindbeck, A., Persson, M.: The gains from pension reform. J. Econ. Literature **41**(1), 74–112 (2003)
61. Milevsky, M.A.: The Calculus of Retirement Income. Cambridge University Press, Cambridge (2006)
62. Modigliani, F., Muralidhar, A.: Rethinking Pension Reform. Cambridge University Press, Cambridge (2005)
63. Musgrave, R.: A reappraisal of financing social security. Public Finance in a Democratic Society **2**, 103–122 (1981)
64. Norberg, R.: Reserves in life and pension insurance. Scand. Actuar. J. **1991**(1), 3–24 (1991)
65. Olivieri, A., Pitacco, E.: Introduction to Insurance Mathematics. Springer, New York (2011)
66. Palmer, E.: The Swedish pension reform model: framework and issues. Social Protection Discussion Papers of The World Bank, p. 12 (2000)
67. Palmer, E.: What is NDC?" In: Pension Reform: Issues and Prospects for Non-Financial Defined Contribution (NDC) Schemes. The World Bank, Washington D.C. (2006)
68. Pena, J.I.: Planes de Prevision Social. Ediciones Piramide, Madrid (2000)
69. Peris-Ortez, M., Alvarez-Garcia, J., Dominguez-Fabian, I., Devolder, P.: Economic Challenges of Pension Systems. Springer, New York (2020)
70. Queisser, M., Whitehouse, E.R.: Neutral or fair? actuarial concepts and pension-system design. OECD Social, Employment and Migration Working Papers **40** (2006)

71. Report, A.: Orange Report 2014 Annual Report of the Swedish Pension System. Technical Report. Swedish Social Insurance Agency, Stockholm, 2014
72. Sakamoto, J.: Japan's Pension Reform. Social Protection Working Papers of The World Bank **541** (2005)
73. Sakamoto, J.: Roles of the Social Security Pension Schemes and the Minimum Benefit Level under the Automatic Balancing Mechanism. NRI Papers **125**, 1–14 (2008)
74. Schokkaert, E., Parijs, P.: Social justice and the reform of Europe's pension systems. J. Eur. Soc. Policy **13**(3), 245–263 (2003)
75. Schokkaert, E., Devolder, P., Hindriks, J., Vandenbroucke, F.: Towards an equitable and sustainable points system. A proposal for pension reform in Belgium. J. Pension Econ. Financ. **19**(1), 49–79 (2018)
76. Selén, J., Ståhlberg, A.C.: Why Sweden's pension reform was able to be successfully implemented. Eur. J. Pol. Econ. **23**(4), 1175–1184 (2007)
77. Settergren, O.: The automatic balance mechanism of the Swedish pension reform. Wirtschaftspolitische Blatter **4**, 1–15 (2001)
78. Settergren, O., Mikula, B.D.: The rate of return of pay-as-you-go pension systems: a more exact consumption-loan model of interest. J. Pension Econ. Financ. **4**, 115–138 (2005)
79. Stevens, Y.: Pension Law. Elgar European Law, Exeter (2024)
80. Subramaniam, I.: Actuarial Mathematics of Social Security Pensions. International Labour Organisation, Geneva (1999)
81. Thornton, P.N., Wilson, A.F.: A realistic approach to pension funding. J. Inst. Actuar. **119**(2), 229–312 (1993)
82. Thullen, P.: Techniques Actuarielles de la Sécurité Sociale. Bureau International du Travail, Genève (1974)
83. Trowbridge, C.L.: Fundamentals of pension funding. Trans. Soc. Actuar. **4**, 17–43 (1952)
84. Turner, J.: Social Security Financing: Automatic Adjustment to Restore Solvency. AARP Public Policy Institute Policy Briefs, Washington, D.C. (2009)
85. Turner, J.: Hybrid Pensions: Risk Sharing Arrangements for Pension Plan Sponsors and Participants. Society of Actuaries, Rosemont (2014)
86. Uebelmesser, S.: Unfunded Pension Systems: Ageing and Migration. Elsevier, Amsterdam (2004)
87. Valdés-Prieto, S.: The Financial Stability of Notional Account Pension. Scand. J. Econ. **102**, 395–417 (2000)
88. Verniere, L.: Panorama des réformes des systèmes de retraite à l'étranger. Questions Retraite. Caisse des dépôts et consignations, Paris (2002)
89. Vidal-Meliá, C., Boado-Penas, M.C.: Compiling the actuarial balance for pay as you go pension systems. Is it better to use the hidden asset or the contribution asset. Appl. Econ. **45**, 1303–1320 (2013)
90. Vidal-Meliá, C., Boado-Penas, M.C., Settergren, O.: Automatic Balance Mechanisms in Pay-As-You-Go Pension Systems. Geneva Papers Risk Insurance Issues Practice **34**, 287–317 (2009)
91. Villalon, J.G.: Manuel de Matematicas Financiero–actuariales. Publicaciones Empresa y Humanidades, Madrid (1994)
92. Viossat, L.C.: Les retraites. Enjeux, crise, solutions. Flammarion. Flammarion, Paris (2000)
93. Weaver, K., Willén, A.: The Swedish pension system after twenty years: mid-course corrections and lessons. OECD J. Budgeting **13**(3), 1–26 (2014)
94. Williamson, J.: Assessing the pension reform potential of a notional defined contribution pillar. Int. Soc. Secur. Rev. **57**(1), 47–64 (2004)
95. Winklevoss, H.E.: Pension Mathematics with Numerical Illustrations. Pension Research Council, Philadelphia (1977)

Index

A

Accrued liability, 195
Actuarial anticipation, 215
Actuarial equilibrium, 61, 176, 183
Actuarial equivalence, 3, 37, 61, 66, 70, 73, 99, 102, 114, 149, 204, 209, 221, 224
Actuarial gains and losses, 229
Adjustment, 129, 169, 187
Adjustment on current pensions, 169
Age coefficient, 181–184, 187
Age pyramid, 7, 8, 45, 48
Aggregate cost, 236
Aging intensity, 53
Amortization period, 242, 245–249
Annuity, 107, 109, 254, 262, 263
Asset gap, 229–231, 234
Attained age normal, 220
Average general premium method, 64
Average of the last few years, 23

B

Back service, 26, 199–202, 204, 207, 220, 226–228
Benefit ratio, 4, 68–70, 72, 92, 149–153, 155–174, 183–188
Beveridgean model, 18
Bismarckian model, 18
Buffer fund, 70, 71, 91, 141, 220

C

Career average, 22, 23, 34, 112, 202
Career length, 25
Cohort, 3, 7, 37, 38, 40, 45, 46, 54, 85, 108, 115, 127, 133, 134, 258, 259
Collective funded method, 191, 192, 194, 196
Constant premium, 198, 199, 211
Contribution rate, 4, 24–28, 56, 57, 61, 63, 64, 67, 68, 71, 72, 74–80, 82, 90–92, 96, 98–101, 105, 112, 113, 123, 149–171, 175, 178, 179, 181, 183, 185–187, 194, 208–210, 219, 224, 226, 227, 236, 237
Conversion factor, 30, 32, 34, 107–109, 113, 119, 121, 183

D

Defined Benefit, 3, 4, 19, 21, 33, 36, 55, 63, 64, 67, 89–91, 95–98, 109, 111–113, 149, 150, 153, 155, 156, 164, 165, 175, 181, 184, 191, 197, 213, 219, 237
Defined Contribution, 3, 4, 19, 21, 33, 35, 36, 63, 90, 91, 95–97, 105, 111, 114, 149, 151, 153, 156, 157, 164, 165, 175, 180, 181, 213
Defined Musgrave, 153, 155, 157, 158
Dependence ratio, 52, 53, 68

E

Entries into the population, 45, 49, 66
Equalization method, 102
Exits from the population, 45

F

Final salary, 3, 23, 24, 31, 32, 55, 76, 91, 116, 164, 202, 203, 213, 215

First pillar, 16–18, 28, 29, 89–91, 93, 96, 97
Frozen initial liability, 220
Fully funded pension, 191

I

Independent probability, 58
Indexed career average, 23
Individual funded method, 192, 194, 196, 197, 199, 215
Individual level premium, 200, 208, 233
Initial funding, 202

L

Level method, 98
Level technique, 98
Liability gap, 230, 233, 236
Life expectancy, 1, 5, 6, 15, 16, 54, 91, 108, 112, 122, 129, 137, 138, 182, 184, 185, 187
Lump sum, 29

M

Macroeconomic indicator, 92
Musgrave ratio, 151, 152

N

Non-stationary equilibrium model, 122
Normal entry age, 200
Normal retirement age, 184
Notional account, 114
Notional capital, 106, 107
Notional rate of return, 106
Number of points, 175–177, 179, 181, 185

P

Pay-As-You-Go method, 40
Pension reform, 95
Pension scheme, 15, 19, 21, 36, 61
Points account, 181, 182
Point system, 97, 175, 180, 181
Portability, 213
Potential support ratio, 53

Price of the point, 175
Projected individual level percent, 200
Projected unit credit cost, 200
Prospective form, 62
Prospective probabilities, 127
Pure endowment, 198, 199, 201, 202, 205, 210, 212, 231, 234, 235, 262, 263

R

Rate of return, 91
Reference career duration, 187
Reference salary, 22
Renewal equation, 46
Replacement rate, 4, 35, 92, 113, 114, 187
Retirement pension, 21, 23, 26, 29, 31, 32, 91, 109, 110, 114, 117–119, 121, 134, 136, 181, 215, 223
Revaluation, 119, 128, 129
Risk community, 17, 37, 38, 40, 41
Risk sharing, 156, 160, 172, 173
Risk-sharing coefficient, 156–160, 162

S

Samuelson–Aaron rule, 81
Second pillar, 16, 17, 19, 20, 28, 29, 66, 94, 219
Social security schemes, 28, 94
Spread method, 237
Survival dividend, 131, 133
Sustainability coefficient, 171–173, 182, 183, 185, 186, 188

T

Third pillar, 16, 94
Turnover duration, 144

U

Unfunded accrued liability, 195
Unit credit cost, 200, 203, 216, 230
Unstable population, 50

V

Value of the point, 176–187

The manufacturer's authorised representative in the EU is Springer Nature Customer Service Centre GmbH, Europaplatz 3, 69115 Heidelberg, Germany. If you have any concerns regarding our products, please contact ProductSafety@springernature.com

Printed and bound by CPI Group (UK) Ltd, Croydon, CR0 4YY

26/03/2026

02078916-0007